Applied Electronic Engineering with

Mathematica®

Applied Electronic Engineering with

Mathematica®

Alfred Riddle

Samuel Dick

Addison-Wesley Publishing Company

Reading, Massachusetts   Menlo Park, California   New York
Don Mills, Ontario   Wokingham, England   Amsterdam   Bonn
Sydney   Singapore   Tokyo   Madrid   San Juan   Milan   Paris

This book was reproduced from text and illustration files supplied by the authors.

Library of Congress Cataloging-in-Publication Data

Riddle, Alfred.
    Applied electronic engineering with *Mathematica* / Alfred Riddle, James Samuel Dick.
        p.   cm.
    Includes bibliographical references (p.   ) and index.
    ISBN 0-201-53477-0
    1. Electronics—Data processing. 2. Mathematica (Computer file).
  I. Dick, James Samuel. II. Title.
  TK7835.R525   1995
  621.3'0285'5369—dc20
                                                93-33040
                                                       CIP

The programs and applications presented in this book have been included for their instructional value. They have been tested with care but are not guaranteed for any particular purpose. The publisher does not offer any warranties or representations, nor does it accept any liabilities with respect to the programs or applications.

Many of the designations used by manufacturers and sellers to distinguish their products are claimed as trademarks. Where those designations appear in this book, and Addison-Wesley was aware of a trademark claim, the designations have been printed in initial caps or all caps.

Mathematica is a registered trademark of Wolfram Research, Inc. *Mathematica* is not associated with Mathematica, Inc., Mathematica Policy Research, Inc., or MathTech, Inc.

package ISBN   0-201-53477-0
book ISBN   0-201-82689-5
disk ISBN   0-201-94252-6

1 2 3 4 5 6 7 8 9 10-CRW-97969594

Preface

The purpose of *Applied Electronic Engineering with Mathematica* is to show how *Mathematica* can solve a wide variety of problems in electronic engineering. The book is intended for a wide audience — undergraduate and graduate-level students in electronic engineering, and instructors, developers, and professional engineers working with computer aided design (CAD). We believe that *Mathematica* is an exciting engineering tool because you can use numerical or symbolic techniques and you also have instant access to powerful tools to aid comprehension of your results: graphics and sound. With *Mathematica*, you can design or analyze a filter circuit, draw a graph of its frequency response, generate some white-noise test data, play the data through the filter, and hear how the filtered output sounds.

You may ask whether *Mathematica* will be a useful tool for you. After all, electrical engineers already have calculators, spreadsheets, programming languages, and many other problem-solving tools. Because *Mathematica* is a *system* for doing mathematics, you will have a complete set of tools for solving your problems. Yes, it takes time to learn *Mathematica*, but we believe that you will save time in the future: When you want to extend the scope of your work or the tools you have designed, you can do so within one environment. *Mathematica*'s environment allows you to use numeric tools, symbolic manipulation, high-level mathematical functions, and graphics without concern for inter-system boundaries, data formats, or other compatibility issues. *Mathematica* is a boon to all engineers. In other words, *Mathematica* allows you to concentrate on the engineering without becoming distracted by the tedium of the mathematical process you are using.

Because *Mathematica* can help you equally well with any level of work, you choose your own level. You can tackle problems ranging from Ohm's law through *s*-domain analysis — from high school problems to research projects at graduate school and in industry. With this range in mind, our goal was to write a book that will be helpful at all these levels. For newcomers to *Mathematica*, we have included throughout each chapter a feature called a Toolbox which briefly describes *Mathematica* functions as they are used. With this learning aid, you can start work without having to know a lot about *Mathematica*. We also describe *Nodal*, an electronic engineering package for use with *Mathematica*. Because *Nodal* examples are described in detail in Section A.2 and elsewhere, the Toolbox sections are limited to *Mathematica* commands.

A demonstration version of *Nodal* is included with the book. This version, **NodalDem**, is limited to four nodes and ten evaluation points but will run all the examples in this book, although, for clarity, the plots in this book typically use more than ten points. All of the code from each chapter is also included on the disk that accompanies the book.

We explain the application of *Mathematica* to engineering problems principally by example. Where we felt appropriate, we have added some description of the underlying engineering, but not with the aim of writing a textbook on the engineering.

In the early chapters, we show you how to use *Mathematica* on simple problems (so that the complexity of the problem does not mask the exposition of the *Mathematica* technique) and, in later chapters, how to program *Mathematica* to tackle complicated problems. Most of our examples are based on analog-circuit design and signal processing. We have tried to ensure that the importance of symbolic analysis is highlighted, especially as a technique that you can use to understand and improve your designs.

Although electrical engineering is mainly concerned with design, most teaching concentrates on analysis. So how can analysis, which may seem rather *post hoc*, help design? We believe that the ability to use symbolic analysis brings you closer to design: You can inspect complicated expressions to watch for significant variables and use derivatives or series expansions to identify sensitive variables. Because *Mathematica* includes symbolics, numerics, graphics, sound, and custom programming, you can view a problem from many different perspectives. We believe that these techniques *can* help you invert analysis into design.

Because *Mathematica* has a programming language, you can automate solutions to your problems. *Mathematica* supports a wide range of programming styles, so you can choose your own style. We emphasize the functional and object-oriented aspects of programming *Mathematica*. Functional programming may seem odd if you were brought up on procedural programming, but it works well for creating reusable code. We will also show some examples of traditional procedural programming for comparison.

The layout of the book has been guided by a philosophy of "simplicity first." Chapter 1 can be browsed through or read thoroughly, depending on your *Mathematica* expertise. We have included Toolboxes to display information on *Mathematica*; experienced *Mathematica* users can skip the Toolboxes. Chapters 2, 3, and 6 solve basic engineering problems with *Mathematica*. Chapters 4, 5, and 7 through 10 delve into computer-aided design and signal processing; they demonstrate functional programming, graphics programming, and package development. These latter chapters will be useful to readers interested in programming *Mathematica*, even if they are not interested in the subject matter. The presentation flows well enough, we believe, that programmers can get beyond the technical details and get a better feel for using *Mathematica*.

We believe that you will find useful the sample programs developed in the chapters and in the demonstration version of *Nodal*. Section A.2 in the Appendix lists the functions within *Nodal*. All of the *Mathematica* code examples which require *Nodal* have a *Nodal* symbol in the right margin (*see* right). (The demon-

N

stration version of *Nodal* omits some of the more advanced functions and is limited to four nodes.) We encourage you to work through the exercises at the end of each chapter; they will allow you to think about how to use *Mathematica* and to test your progress. We both enjoy teaching and have found James 1:22 good advice. Finally, this book is not a substitute for Wolfram's book, *Mathematica*, or Maeder's book, *Programming in Mathematica*. If you decide to create packages for others to use, the information on Contexts and Packages in Maeder's book will be invaluable.

Acknowledgments

We thank the many people who have helped us write this book: Stephen Wolfram, at Wolfram Research; Kevin McIsaac and Roman Maeder, formerly of Wolfram Research; and the staff at Addison-Wesley, including our publishers, Peter Gordon and Helen Goldstein, our editor, Lyn Dupré, and Juliet Silveri, in production. A special thanks is reserved for Allan Wylde, who is now at Telos but who started this book at Addison-Wesley. We would also like to thank Glenda and Alison, without whose patience and support the project would not have been possible, and a special thanks goes to Puzon, Boson, Locai, and Glover, whose understanding of and perspective on life helped restore calm during the frustrations and turbulence of writing this book.

Alfred Riddle
Samuel Dick

The Nodal Package and Disk

The 3.5-inch disk accompanying this book includes a demonstration version of *Nodal*, and the *Mathematica* code from each chapter. The disk is in a DOS format, which almost all current computer systems can read. You will, of course, need *Mathematica* to run the software on the disk. *Nodal* and the *Mathematica* code in this book are compatible with *Mathematica* versions greater than 2.0 and will run on any machine that runs *Mathematica*.

The *Mathematica* code from each chapter is an ASCII text file: any word processor can read the text files. When the word processor reads the files, the only problem it might have is with line termination characters. Different computer systems use different line termination characters: DOS systems use carriage returns and linefeeds; Macintosh computers use carriage returns only; and UNIX systems use linefeeds only. You may need to remove one of the line termination characters so that your word processor, or even *Mathematica*, can read the files properly. The *Mathematica* Notebook front ends should handle any combination of line termination characters — but some versions do not, so you should beware. The *Nodal* package is encoded. We have included all the functions and utilities of *Nodal* on the disk because we felt the readers of this book would want to try all of *Nodal*. The only limitations we have imposed are limiting the plots to ten points and limiting the nodes to four. The commercial version of *Nodal* is not encoded and contains no node or datapoint limitations. Because the commercial version of *Nodal* is in ASCII text, users can add their own models to *Nodal*. Working with *Mathematica* allows us to develop inexpensive and powerful software. Yet, working on top of *Mathematica* makes it difficult for us to deliver a useful demonstration version of *Nodal* without giving away our software. We hope you find the demonstration version of *Nodal* a good compromise. Please read the READ-ME.TXT file on the disk for further information.

Nodal was written by one of the authors, Alfred Riddle of Macallan Consulting. Information, pricing, and help with *Nodal* is available from:

Macallan Consulting
1583 Pinewood Way
Milpitas, CA 95035
USA

Contents

Preface   *v*
    Acknowledgments   *vii*
    The *Nodal* Package and Disk   *vii*

1 Groundwork   *1*

1.1 The Symbolic Advantage   *1*

1.2 Computer-Aided Engineering   *2*
    1.2.1 How Do You Want CAD to Help You?   *3*
    1.2.2 *Mathematica* and *Nodal*   *5*

1.3 First Steps   *5*

1.4 Foundation Skills   *6*
    1.4.1 Basic Electronics Knowledge   *6*
    1.4.2 A Little *Mathematica*   *6*
    1.4.3 Differences Between Linear and Nonlinear Circuit Analysis   *7*
    1.4.4 Netlists   *7*
    1.4.5 Value Scaling with Qualifiers   *11*
    1.4.6 Naming of Objects   *14*
    1.4.7 On-Line Help   *14*
    1.4.8 Options   *16*
    1.4.9 Layout and Style   *17*

1.5 Error Messages   *18*

1.6 Summary   *19*

1.7 References   *19*

2 DC Circuits   *21*

2.1 Voltage and Current in Resistive Circuits   *21*
    2.1.1 Mesh Analysis   *26*
    2.1.2 Multistate DC Circuits   *33*

2.2 Power-Conversion Utilities   *36*

2.3 Attenuator Design   *41*

2.4 Nonlinear DC Circuits   *43*

    2.4.1 Device Characterization Using Experimental Data   *44*

    2.4.2 Nonlinear Circuit Analysis   *52*

2.5 Summary   *59*

2.6 Exercises   *59*

2.7 References   *60*

3 Small-Signal Circuits I: Introduction   *61*

3.1 Magnitude and Phase Calculation in RLC Circuits   *61*

    3.1.1 Elementary Technique   *61*

    3.1.2 Mesh Analysis   *68*

    3.1.3 Nodal Analysis   *71*

3.2 Small-Signal Analysis Techniques   *74*

    3.2.1 Preparation for Analysis   *74*

    3.2.2 Dynamic and Static Device Properties   *76*

3.3 Analysis of Circuits Containing Active Devices   *79*

    3.3.1 Voltage-Controlled Voltage Sources   *79*

    3.3.2 Operational Amplifiers   *83*

    3.3.3 Voltage-Controlled Current Sources   *85*

    3.3.4 Current-Controlled Current Sources   *89*

    3.3.5 Current-Controlled Voltage Sources   *89*

3.4 Device-Equivalent Circuits   *90*

    3.4.1 The FET Model   *90*

    3.4.2 The BJT Model   *91*

3.5 Oscillators and Feedback Network Design   *92*

3.6 Summary   *97*

3.7 Exercises   *97*

3.8 References   *98*

4 Small-Signal Circuits II: Multiport Analysis   *99*

4.1 Introduction to the Analysis Method   *100*

    4.1.1 Y-Parameters   *100*

    4.1.2 Z-Parameters   *101*

4.1.3 ABCD-Parameters   *102*
4.1.4 Parameter Calculation   *102*

4.2 Matrix Analysis with *Mathematica*   *105*

4.3 General Matrix Analysis   *112*
4.3.1 *ABCD*-Matrices   *113*

4.4 Matrix Transformations   *115*
4.4.1 Object-Oriented Technique   *116*

4.5 CAD Program   *117*

4.6 S-Parameters   *121*
4.6.1 Reflections and Matched Terminations   *122*
4.6.2 Use of S-Parameters in Analysis   *123*
4.6.3 Conversion of Z-Parameters to S-Parameters   *127*
4.6.4 Properties of S-Parameters   *128*

4.7 Summary   *131*

4.8 Exercises   *131*

4.9 References   *132*

5   Component Design and Sensitivity Analysis   *133*

5.1 Component Value Functions and Utilities   *133*

5.2 RLC Filter Design   *134*

5.3 What-If Sensitivity Analysis   *138*

5.4 Differential Sensitivity Analysis   *145*

5.5 Cost Minimization   *148*

5.6 Summary   *150*

5.7 Exercises   *151*

6   Time Series and Spectral Analysis   *153*

6.1 Time Series   *153*
6.1.1 Generation of Time Series   *153*
6.1.2 Statistical Analysis and Plotting   *154*
6.1.3 Generation of Time Series with Specific Noise Properties   *157*

6.1.4 Correlation and Convolution   *160*

6.1.5 Synthesis of Functions   *163*

6.2 Fourier Analysis   *165*

6.3 Frequency-Domain Filtering   *167*

6.4 Summary   *169*

6.5 Exercises   *170*

6.6 References   *170*

7   *s*-Domain (Laplace) Analysis   *171*

7.1 Laplacian Description of Signals   *171*

7.2 *s*-Domain Transfer Functions   *173*

7.3 Visualization of Pole-Zero Descriptions   *174*

7.4 Determination of Circuit Impulse and Step Responses   *176*

7.5 Summary   *186*

7.6 Exercises   *186*

7.7 References   *187*

8   Filter Design   *189*

8.1 Transfer Functions   *189*

8.1.1 The Butterworth Response   *191*

8.1.2 The Chebyshev Response   *192*

8.1.3 Pole-Zero Locations   *193*

8.1.4 Component Values   *196*

8.2 Transformations   *199*

8.2.1 Impedance Scaling   *199*

8.2.2 Frequency Scaling   *201*

8.2.3 High-Pass Transformation   *204*

8.2.4 Band-Pass Transformation   *206*

8.3 Basic Synthesis   *210*

8.3.1 Singly Terminated Synthesis   *211*

8.3.2 Component Values by Continued Fraction Expansion   *211*

8.4 Advanced Synthesis   *215*

   8.4.1 Doubly Terminated Synthesis   *216*
   8.4.2 Spectral Factorization   *216*
   8.4.3 A Butterworth Example   *219*
   8.4.4 Pole-Zero Extraction   *220*
   8.4.5 Design by Optimization   *224*

8.5 Digital Filtering   *230*

   8.5.1 Sampling of Signals   *231*
   8.5.2 Mapping of *s*- to *z*-Plane   *233*
   8.5.3 Infinite Impulse Response Filters   *233*
   8.5.4 Finite Impulse Response Filters   *238*

8.6 Summary   *241*

8.7 Exercises   *241*

8.8 References   *242*

9   High-Frequency Circuits and Analysis   *245*

9.1 The Smith Chart   *245*

   9.1.1 Impedance and Reflection   *245*
   9.1.2 Generation of a Smith Chart   *248*
   9.1.3 Smith Chart Function   *250*

9.2 Stability Analysis Using S-Parameters   *254*

   9.2.1 Device *K* Factor   *255*
   9.2.2 CAD and the Stability Factor   *256*

9.3 Stability Circles   *258*

   9.3.1 Gain Circles   *260*
   9.3.2 A Gain Circle Example   *263*

9.4 Matching-Network Design   *265*

   9.4.1 Smith Chart Impedance Traces   *265*
   9.4.2 Smith Chart Admittance Traces   *267*
   9.4.3 Matching-Network Design   *268*
   9.4.4 Design Evaluation with *Nodal*   *271*

9.5 System Design   *273*

   9.5.1 Cascade Analysis Mathematics   *274*
   9.5.2 A Cascade-Analysis Program   *276*
   9.5.3 Plotting the Cascade Analysis   *279*
   9.5.4 Drawing the Cascade   *280*

9.6 Summary   *281*

9.7 Exercises   *282*

9.8 References   *283*

10 Noise Analysis   *285*

10.1 Random Signals   *285*

    10.1.1 White Noise   *286*
    10.1.2 Brown Noise   *290*
    10.1.3 Pink Noise   *292*

10.2 Autocorrelation and Power   *295*

10.3 Multiple Signals and Correlation Matrices   *299*

    10.3.1 The Correlation Matrix   *300*
    10.3.2 Resistor Noise   *300*
    10.3.3 Circuit Noise   *301*

10.4 Noise Matrix Analysis   *304*

    10.4.1 Noise Matrix Description   *304*
    10.4.2 Converting Correlation Z-Matrix to Y-Matrix   *305*
    10.4.3 The Correlation $ABCD$-Matrix   *308*

10.5 Noise Figure   *310*

    10.5.1 Signal-to-Noise Degradation   *310*
    10.5.2 Noise Figure and Noise Sources   *311*
    10.5.3 Noise Figure Relationships   *313*

10.6 Noise Solutions with *Nodal*   *315*

10.7 Summary   *317*

10.8 Exercises   *317*

10.9 References   *318*

Appendix   *319*

A.1 *Mathematica* Functions   *319*

    A.1.1 Syntax   *319*
    A.1.2 Numbers   *320*
    A.1.3 Lists   *320*
    A.1.4 Manipulating Lists   *320*

A.1.5   Basic Arithmetic Operations   *321*

A.1.6   Complex Numbers   *321*

A.1.7   Rule Symbol and Replacement   *321*

A.1.8   Manipulating Expressions   *321*

A.1.9   Prefix, Infix, and Postfix Forms of Operators   *322*

A.1.10   Matrix Multiplication   *322*

A.1.11   User-Defined Functions   *322*

A.1.12   Reading Data from ASCII Files   *322*

A.1.13   Fitting Data to Functions   *322*

A.1.14   Manipulating Equations   *322*

A.1.15   Calculus   *323*

A.1.16   Anonymous Functions   *323*

A.1.17   Random Number Generation   *323*

A.1.18   Laplace Transform   *324*

A.1.19   Manipulating Polynomials   *324*

A.2   *Nodal* Components, Functions, Utilities, and Constants   *324*

A.2.1   Components   *324*

A.2.2   *Nodal* Functions   *331*

A.2.3   *Nodal* Utilities   *333*

A.3   Graphics   *340*

A.3.1   Lists   *340*

A.3.2   Log Plots   *343*

A.3.3   Adding a Legend   *344*

A.3.4   Multiline Plots   *345*

A.3.5   Magnitude and Phase Plots   *346*

A.3.6   Polar Plots   *348*

A.3.7   Smith Charts   *351*

A.4   Importing Data   *351*

A.5   Exporting Data   *357*

A.6   Example Code Usage   *359*

A.7   Complex Algebra   *366*

Index   *371*

CHAPTER 1

Groundwork

Using *Mathematica*, you can carry out many computations in electrical and electronic engineering symbolically as well as numerically — from simple DC circuit analysis to microwave impedance matching and Laplacian analysis.

What does this ability mean? Why is it an advantage?

1.1 The Symbolic Advantage

If you analyze a circuit and wish to know, say, the input impedance, then, given all the component values in the circuit, *Mathematica* is able to provide a numerical answer. That is numerical analysis. But you could also choose to leave out one or more component values — referring to omitted values by symbolic names such as "*rFeedback*." Analyzing the circuit symbolically, *Mathematica* is able to give you an equation, in *rFeedback*, for the input impedance. For example, the two equations

$$Z_{in} = 1300 \ \Omega \text{ and}$$

$$Z_{in} = 14/rFeedback \ \Omega$$

both tell you about the input impedance, Z_{in}. The first represents a purely numerical result. The second shows a *symbolic* result and is much more informative: Not only can it tell you what value the impedance is (given a value for *rFeedback*) but it can also tell you how the impedance will be affected by *rFeedback*. What happens when *rFeedback* becomes small? Which other components affect the input impedance? The symbolic form answers these questions instantly.

Hence, symbolic circuit analysis allows you to keep problems general and often to gain more insight into the circuits that you design. With the symbolic answer, if *rFeedback* changes, you do not have to recompute the symbolic answer — you merely have to reevaluate it.

Of course, it is possible to rerun an all-numerical analysis with different values of *rFeedback* and to note what happens. You could then use the fitting functions in *Mathematica* to fit an analytical function to the results. However, we hope that you agree that the symbolic answer is simpler and more informative than is the numeric answer. Symbolic circuit analysis is just one of the aspects of applying *Mathematica* to electronic engineering that we explore in this book.

Let us begin.

1

1.2 Computer-Aided Engineering

This book is about *computer-aided electronic engineering*. In particular, it is about making the most of a computer as a tool. Many engineering tasks are so complicated that computers must be used to solve them. Indeed, many design problems have so many different aspects that no single computer program can solve the entire problem, and engineers usually resort to a combination of analysis, experience, and rough calculations.

Tools such as *Mathematica* offer a nearly ideal foundation for engineering problem solving. Because *Mathematica* is a complete system for mathematics on the computer, it covers many aspects of engineering design. You can perform rough calculations, generate graphics, and compute detailed analyses using compiled programs. To date, the only facility missing from *Mathematica* is a graphical front end to support circuit schematics and sketches. A major compensation for this missing facility is the Notebook front end for *Mathematica*, which is currently available on the Apple Macintosh, NeXT, and IBM personal computers running Windows. The *Mathematica* Notebook provides a single document that contains each analysis in a cell. These cells can be grouped with other cells and can be hidden if desired. So, you can construct a readable and effective laboratory notebook with each large grouping of cells representing, and containing, another aspect of the problem to be solved. Your *Mathematica* Notebook is easily transferred to colleagues for verification or documentation, too.

For a computer to be useful, it must be available and have the software to solve the problem at hand. Although computer time used to be expensive, relatively powerful personal computers and workstations are now on the desks of most engineers: Computers are available and time on them is cheap. The problem of having the right software is not so easily dismissed.

Software usually comes from one of two sources: home brew or commercially produced. In the past, the time required to write computer programs was long (still true for some computer languages), whereas commercially available computer programs were, and still are, often expensive and could be hard to use — either because they are difficult to learn or because they do not work exactly as you want. Therefore, many technical computer users undertake programming to provide themselves with the software they require.

Programming tools have also improved (compare early FORTRAN with the sophistication of C++ or Ada) but, as an engineer, you have special needs — a language that supports complex numbers, matrices, and graphics is highly advantageous. You also want an interactive language so that you can try out your ideas quickly. Finally, you need to save your work in functions that are easily recalled by you or your colleagues.

Although applications of symbolic mathematics on computers are not new [Nolan53] they are still not widespread, so most engineers focus on only numerical applications of computers. Because many engineering calculations are algebraic and great emphasis is placed on formulae, symbolic manipulation is important — especially in design, synthesis, and optimization: Symbolic manipulation should be an important *sine qua non* for any new computing tool.

The purpose of this book is to help engineers make the most of their computer. We believe *Mathematica* has a major role to play because it offers a mathematical environment and is available on a wide range of computers. *Mathematica* combines (complex) numerical facilities, matrices, graphics, programming, and symbolic manipulation in a truly interactive way. We demonstrate the application of *Mathematica* to electronic engineering and present many algorithms and functions to solve specific problems.

Although *Mathematica* is capable of complicated circuit analysis operations, we always try to illustrate its use using simple circuits. The reason we use simple circuits is not that *Mathematica* cannot cope with more complex examples — it is just easier to concentrate on learning how to use *Mathematica* if you are not distracted by the complexities of the examples. So this strategy forms the basis of the book's approach to the subject.

The contents of this book are merely a subset of how *Mathematica* may be able to help you: One of the advantages of computer technology is that the only real limit on how the technology can help you is human imagination.

Of course, there are practical limits to certain ways of working, but it is almost always possible to enlist the computer's help along some part of your journey toward a solution. It is important to remember that computer-aided design is just that. The computer will not design the equipment for you. Just like when you set off on a journey, you must know where you are, know where you want to go, and have some idea about which route and which mode of transport you intend to use. Bearing these considerations in mind, you will find that *Mathematica* is able to help you in a considerable number of ways. Its ability to handle engineering problems using symbolic descriptions is a powerful asset.

1.2.1 How Do You Want CAD to Help You?

Can Mathematica help me check my design?

Given a description of your design, *Mathematica* will be able to confirm your calculations. You can use *Mathematica* to perform or help check algebraic manipulation or to perform the evaluation of circuit characteristics over a wide range of parameters — in general, you can use *Mathematica* with more ease than you could a good pocket calculator.

Will Mathematica allow me to try out different circuit configurations?

With the ease of cut and paste, you can try different configurations as well as use different components — all without having to use either a push-in/pull-out breadboard or a soldering iron.

By comparing the CAD-derived answers with your requirements, you can identify quickly how well each configuration satisfies your needs. You often will be able to identify which circuit parameters drive circuit performance more easily using symbolic analysis.

But is the CAD answer too ideal?

With CAD, you are able to try out circuits at varying levels of realism. You can add stray capacitances, inductances of "noninductive" components, and small resistances that will be present in the tracks of printed circuit boards. The complexity of your CAD model will be reflected in the realism of the modeled circuit's parameters. Whether you want a quick look or an in-depth analysis is entirely up to you.

I am just beginning to do circuit analysis — is CAD too complex for me?

Learning about any new subject takes time. There will always be parts of the learning process that require more study than do others. By using *Mathematica*, you can avoid much of the mathematical tedium, so you have more time to concentrate on the fundamentals of the subject. Just as learning how to use a scientific calculator takes some time, so *Mathematica* requires time, too, but the payback time is short, and after that you are able to save time consistently — for many years to come.

How much computer knowledge will I need?

You will need to know enough to switch on the computer, and to start up and use an applications program. That is all you need to know.

But CAD will not design a circuit for me, right?

CAD can help you during design. By using symbolic analysis, you can find out how circuit components affect the required performance, and that knowledge can point you toward better design — and help you to avoid pitfalls. You can try out ideas using the CAD system as an impartial judge. When a design is not working, you can look at, say, voltages around the circuit to see whether feedback is right or wrong or whether a signal is being short-circuited. At the end of the day, CAD will not take over your task of design — although it may proffer a second opinion.

At what stage of circuit development should I use CAD?

That is up to you. Many people prefer to use CAD as a helping hand from the beginning; other people prefer to use it merely to check their designs.

Is it easier just to use a numerical package?

Mathematica can act as a numerical tool, but its symbolic abilities are there, in parallel and when you want them, to help. Engineering is a practical subject and will almost always have real-world numbers as its product — a resistor is 4 mm long and has a value of 2000 Ω — but good engineering has a backbone of mathematics with which symbolic CAD can help.

1.2.2 *Mathematica* and *Nodal*

Much of this book is about using *Mathematica* in an electronic engineering context. We also describe the use of *Nodal,* a symbolic AC circuit analysis package designed to be used with *Mathematica*; a version of *Nodal* is supplied on disk with this book. *Nodal* (which is written in *Mathematica*) provides *Mathematica* with a set of rules that teaches *Mathematica* about electronic circuit analysis; it also contains many useful utilities and predefined constants.

  Mathematica applies mathematical rules to input from the user when performing mathematical manipulation. For example, if you want to integrate a function, you need to know the rules by which such an operation can be done — using *Mathematica* to do this task would merely involve using the function **Integrate** because *Mathematica* already contains many rules for integration. *Mathematica* does not, however, contain rules and techniques for circuit analysis — it uses *Nodal* to extend its knowledge in that area. Where we have used a function from *Nodal,* we have placed an N in the right-hand margin.

N

1.3 First Steps

If *Mathematica* is not installed on your machine, then you must install it now. Each copy of *Mathematica* comes with installation instructions. You may also want to increase the amount of memory that the computer allocates to *Mathematica*. Because both installation and memory allocation are machine specific, you will need to refer to the *User's Guide* that came with your copy of *Mathematica*.

  The circuit analysis package *Nodal* is supplied as a set of files that contain the rule base for *Mathematica*. These files must be copied into the *Mathematica* directory on the computer, as described in the installation instructions in *Nodal*'s manual.

  Once you have activated *Mathematica* and it has completed its start-up work, you can load *Nodal* into *Mathematica* using *Mathematica*'s **Needs** command or the `Get` (`<<`) command, as described in this book and in the *Nodal* manual. (Loading may take a few minutes: *Mathematica* has to read in the files and to interpret the rules and functions within the files.) You must load *Nodal* before you can use any of the functions, components, or constants it contains. After loading *Nodal*, *Mathematica* can be used in the normal way; the effect of loading *Nodal* is to extend *Mathematica*'s command repertoire.

On certain computers, you can shorten start-up time of future sessions by saving a core image of the file on which you are working using *Mathematica*'s `Dump` command — the resulting file occupies several megabytes of disk space. However, if you have saved a core image, then, once the computer has loaded the `Dump`ed image, you can begin work exactly where you left off. Thus, you will avoid having to load *Mathematica* and any packages or function definitions that you may have written and been using.

1.4 Foundation Skills

Throughout the book, we introduce you to features of *Mathematica* when you need to apply them. However, if you have certain skills *ab initio,* then you will surely make quicker progress. In this section, we state what knowledge we assume you have in electronics and introduce you to some *Mathematica* concepts and basic operating knowledge that you will find helpful.

1.4.1 Basic Electronics Knowledge

For the early chapters, we assume that you are familiar with applying Ohm's law and, for AC circuits, with handling complex numbers (that represent the magnitude and phase of voltage and currents in resistive and reactive circuits). We also assume that you are familiar with the concepts involved with Kirchhoff's laws — that the net voltage around a loop in a circuit is zero, and that the sum of currents into and out of a circuit node, is zero. You will be able to solve a wide range of circuit problems using only this knowledge and *Mathematica*.

When you work with later chapters, a knowledge of calculus, Fourier and Laplace techniques, and signal-processing concepts will be helpful.

1.4.2 A Little *Mathematica*

If *Mathematica* is new to you, then we strongly advise you to explore it before proceeding further; you do not have to be especially familiar with *Mathematica* to benefit from it, and we explain how to use *Mathematica* commands as they occur in this book. The first few chapters of Stephen Wolfram's book *Mathematica — A System for Doing Mathematics by Computer* [Wolfram91], which comes with the *Mathematica* software, provides a good introduction to the software's many capabilities, and we recommend that you at least browse through those introductory chapters. (*Mathematica* and *Nodal* share the same command syntax, so, once you are familiar with *Mathematica*, using *Nodal* incurs little extra learning overhead.)

Because this book is about using *Mathematica*, we have made every attempt to use *Mathematica* notation and to display equations in ways native to *Mathematica*. Input to *Mathematica* (that is, what you type in) is in a boldface **Courier** font, preceded by *In:*, whereas output generated by *Mathematica* is set in regular Courier, preceded by *Out:*; we have left output in *Mathematica*'s default output format.

Mathematica is a rich mathematical environment and contains many functions — tools — that you will find useful. To learn them all first, before applying them to your engineering problems, may not be the best use of your time. Mostly, you will be able to pick up tools as you go along because, each time we introduce a new *Mathematica* tool, we use a feature we call Toolbox to describe it and explain how you can apply it. The Toolbox text is not a replacement for the full description of *Mathematica* functions that appear in Stephen Wolfram's book *Mathematica — a System for Doing Mathematics by Computer*, but we hope you will find it useful. Even if you are disinterested in some chapters, you may wish to browse through the Toolbox features they contain.

1.4.3 Differences Between Linear and Nonlinear Circuit Analysis

Using *Mathematica*, you can solve many circuit problems, such as calculating currents and voltages in networks, even in simple nonlinear circuits, such as might be involved in the calculation of semiconductor bias points. *Nodal* is able to help you with certain of these problems, too, but it is principally an AC small-signal analysis tool.

The difference between linear and nonlinear circuit analysis techniques is important. For example, in an amplifier, *Nodal* can compute the gain. Say the gain is 100. Applying a 10 µV signal to the input will result in a 1 mV output signal. Applying a 1 V signal to the input will not produce a 100 V output, however, because the amplifier runs from a 9 V battery and will be completely saturated by such a large input signal. Hence, the unfortunate amplifier will have been forced into a nonlinear region of operation; *Nodal* version 2.0 does not perform analysis of circuits operated in nonlinear regions.

1.4.4 Netlists

To use *Nodal*, you must be able to tell *Nodal* what components are in the circuit and how they are interconnected. The traditional way to communicate this information is to use a netlist. A *netlist* is simply a list of components that states what each component is, what its value is, and to which other components it is connected. The netlist uses the concepts of nodes, the junction points between components. For example, Figure 1.1 shows a circuit with various components and the node numbers.

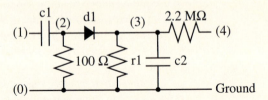

Figure 1.1 A simple circuit showing the location and identification numbers of nodes.

Because a circuit is described by nodes and their interconnecting components, it is often called a *nodal network*. *Nodal* uses this term as the name for a function that takes the netlist as its argument and stores the information about the network you have defined for *Nodal*'s use.

Toolbox

General Command Syntax

Commands and functions have names that begin with an upper-case character. Arguments are separated by commas and enclosed in square brackets—**Command**[*arg1*, *arg2*], for example.

Lists

A list is a collection of items (for example, numbers, symbols, or other lists) enclosed in curly brackets: {*item1*, *item2*, *item3*}. You can assign names to lists, just like any other variable and then refer to the whole list by using its name. Some functions take arguments that are lists.

You can refer to the *i*th member of a list named **myList** by **myList[[i]]**. *Mathematica* uses the double square bracket to distinguish between a list member reference and a function call. (If you used **myList[i]**, *Mathematica* would interpret this as a call to a function named **myList** with a single argument called **i**.)

```
In:
  myList={2,3,a,x^2}
```

```
Out:
  {2, 3, a, x }

In:
  myList^2
 Out:
            2   4
  {4, 9, a , x }

In:
  myList[[3]]
Out:
  a
```

Here is how the circuit in Figure 1.1 is described to *Nodal*:

```
In:
  Needs["Nodal2`"];
  NodalNetwork[Capacitor[{1,2},c1],
               Resistor[{2,0},100],
               Diode[{2,3},100,1 pF,100],
               Resistor[{3,0},r1],
               Resistor[{3,4},2.2 MOhm],
               Capacitor[{3,0},c2]
               ]
```

The command `Needs["Nodal2`"]` causes *Mathematica* to load the package
`Nodal2`; once *Nodal* is loaded, *Mathematica* will be able to interpret the nodal
functions that follow. The command `NodalNetwork[`*arguments*`]` instructs
Nodal to interpret *arguments* as a netlist. Note that, in common with *Mathematica*'s commands, *Nodal*'s begin with an upper-case letter and enclose their arguments in square brackets.

The arguments to `NodalNetwork` are all electrical components from
Nodal's library. Each component in *Nodal*'s library uses the same generic syntax:
`ComponentName[`{*nodes*}`, `*value*`, `*options*`]`.

The nodes to which a component is connected are enclosed in curly brackets
(so forming a list). The nodelist is always the first argument. Component values
and options (if any) follow the nodelist, separated by commas; optional arguments can be omitted. The `Resistor[]` component is shown with two arguments: a list of the nodes between which the resistor is connected `{1,2}` and the
value in ohms, `100`.

You must allocate node identification numbers without gaps: If you have a
node 6, for example, then you must also have nodes 0, 1, 2, 3, 4, and 5 allocated.
Node 0 must be assigned and must be the ground node. If nodes have not been
used contiguously, *Nodal* will warn you with the messages:

Out:
```
NetworkNodes::nodeOrder:
The nodes used in the network are not in order, or the
ground (0) node has been left out.

Missing Nodes = {4}
```

Returning to the example shown in Figure 1.1, we have given some of the components symbolic names, such as **c1**, rather than specific values such as 100 µF. Names that you define — for whole circuits or for single components — should begin with a lower-case letter to prevent confusion with built-in *Mathematica* functions and variable names; you must not assign the single character variables **f**, **p**, **s**, and **c**, because these names have a special meaning to *Nodal* (for example, **c** is equated to the speed of light; we discuss the other variables when they are used).

You can also give components names and values.

Toolbox

Assigning and Deassigning Values to Variables

To assign (Set) a value to a variable: *variableName* **=** *value*

To deassign (Unset) a value to a variable: *variableName* **=.**

For example, you can assign a value of 100 Ω to the resistor referred to as **r1** in the preceding example by typing

In:
```
r1=100
```

In any subsequent analysis, *Mathematica* will assume that you want the numerical value of the variable to be used, rather than the symbolic name. After you have assigned or changed values in a network, you must rerun the **NodalNetwork** command on the network so that *Nodal* is aware of the changes you have made; any analysis carried out on the network before you rerun the **NodalNetwork** command will use previously assigned values (if any). If there was no previous numerical assignment, then the symbolic name will be used.

If you wish to clear a value and to return to the symbolic use of a variable, then use the **Unset** (**=.**) command; for example, to remove the value of 100 assigned to **r1**, type **r1=.** or **r1=270** if you want to change the value from 100 to 270.

If you wish to assign the same numerical value to several variables, you can do so in one line:

```
In:
  r1=r2=r3=100
```

1.4.5 Value Scaling with Qualifiers

You may find it easier always to work in SI units such as ohms, farads, meters, and henries. *Nodal* always works in SI units. For smaller or larger values, *qualifiers* are useful to help scale quantities, thus making equations and variables easier to read. For example, **kOhm** and **MOhm** are good names for qualifiers to provide the required scaling for resistors.

Toolbox

Basic Arithmetic Operations

The operators **+** and **–** have their usual meaning. *Mathematica* uses a space or the asterisk (*****) character as the multiplication operator. Division is accomplished using the slash (**/**) character.

Note that there must always be a space (or an asterisk) between the value and the qualifier, because *Mathematica* treats the space as a multiplication, just as in mathematics it is assumed that **A B** means "**A** times **B**"; **AB** (without a space) will be treated as a variable called "**AB**." You could, of course, always replace the space by *****, to make the multiplication operator more obvious, because *Mathematica* treats a space and an asterisk as synonymous.

Defining your own qualifiers is simple: Just choose a name and assign a value to it. For example, if you have times that are in minutes and you have decided to work in seconds, then

```
In:
  min = 60
```

will define a qualifier called **min** with the value of 60. So when you want a time that is in minutes to be used, say, as an argument to a function that expects seconds, you can qualify the argument value with **min**, thus converting the argument from minutes to seconds. For example, what is the interval, in seconds, between three and five minutes?

In:
```
5 min - 3 min
```

Out:
```
120
```

Division by **min** would achieve the inverse conversion (that is, from seconds to minutes).

Nodal provides the following qualifiers to scale quantities:

```
cm, mm, micron, um
mF, uF, nF, pF, fF
mH, uH, nH, pH, fH
mSec, uSec, nSec, pSec, fSec
kHz, MHz, GHz, THz
kOhm, MOhm
kV, mV, uV, nV, pV, fV
mA, uA, nA, pA, fA
MW, kW, mW, uW, nW, pW, fW
```

For other powers of 10, you must use explicit exponentiation.

Toolbox

Exponentiation

Mathematica uses the ^ symbol for exponentiation: **10^7** is equal to ten million. To raise *e* to a power, you can use **Exp[***power***]**.

So, **14 GHz** is equivalent to

```
14 10^9 or
14*10^9 or
14000000000
```

Beware of using other forms of scientific notation. Although many computer languages support notation such as **1.2e4** to mean "1.2*10^4", *Mathematica* will consider this to mean "1.2 times the variable called e4":

In:
```
1.2e4
```

Out:
```
1.2 e4
```

Similarly with a FORTRAN-like f (floating) number,

> *In:*
> **1.4f4**
>
> *Out:*
> 1.4 f4

Instead, use

> *In:*
> **1.2 10^4**
>
> *Out:*
> 12000.

If you mean "1.2 times *e* (2.818) raised to the fourth power," then you use the **Exp** function.

> *In:*
> **1.2 Exp[4]**
>
> *Out:*
> $$1.2 \; E^4$$

Or

> *In:*
> **1.2 E^4**
>
> *Out:*
> $$1.2 \; E^4$$

Because *Nodal* returns values in SI units, you may find it better to divide (**/**) the result of a calculation by a qualifier, thus converting the answer into a more convenient form, by using the general syntax:

> **value = Function[argument] / qualifier**

For example, to calculate **c1** in picofarads, rather than in farads, use

> **c1 = Capacitance[...]/pF**

The *Mathematica* package **Miscellaneous`Units`** contains many conversion factors. *Nodal* also contains a few conversion factors, especially intended to be of use for microstrip and cable calculations. Using them, you can convert to meters mil (0.001 inch), inches, and feet:

> *In:*
> **5 mil**
>
> *Out:*
> 0.000127

```
In:
  2 feet
```

```
Out:
  0.609601
```

```
In:
  36 inch
```

```
Out:
  0.914402
```

1.4.6 Naming of Objects

You can assign a name to an object — to a component, circuit, or result. Names can be any length. You use the same syntax as you use to assign a value to a variable. For example, to assign a name to a circuit, use

```
name = NodalNetwork[...]
```

You must not redefine the single character names **f**, **p**, **s**, and **c**, because these variable names have special meaning to *Nodal*. For example, **f** is used as *Nodal*'s variable for frequency and is often referred to in analysis; you can assign values to **f** or use it as the frequency variable in transfer functions that you define.

In computer programming, it is common for the underscore character to be used to split up concatenated names to improve their readability: "beam_magnet_current" is easier to comprehend than is "beammagnetcurrent." However, the underscore character has a special meaning in *Mathematica* and so must not be used in this fashion. (For example, **x_Complex** specifies **x** to be any complex value.) An alternative format for improving readability is "beamMagnetCurrent."

1.4.7 On-Line Help

If you cannot remember the attributes of a command, utility, or component, you can use the *on-line help* by typing **?** before the name:

```
In:
  ?Sin
```

```
Out:
  Sin[z] gives the sine of z.
```

This form of requesting help also works for *Nodal*'s commands:

```
In:
  ?Diode
```

N

Out:
```
Diode[{node1,node2},Rd,Cd,Rs,options] specifies a
diode for the NodalNetwork. Rd and Cd are the junction
dynamic resistance and capacitance, respectively. Rs
is the parasitic bulk and contact resistance. Options
include the diode temperature, with Temperature->300
(Kelvin) as the default.
```

When you use the **?** command, *Mathematica* will also tell you what values have been assigned to a name. If a variable has been declared but no assignment made, *Mathematica* will repeat the name:

In:
```
?r12
```

Out:
```
r12
```

But if no definition or assignment has been made, *Mathematica* issues a warning message:

Out:
```
Information::notfound: Symbol r1 not found.
```

Where a value has been assigned, that value is shown:

In:
```
r2=40000
?r2
```

Out:
```
r2 = 40000
```

If you forget the name of a function or component and you are using a Notebook interface, then *Mathematica* can help you to find the name. Type the first few characters of the name, pull down the Action menu, and select the Prepare Input option. Choosing Complete Selection from the submenu will produce a list of both *Nodal* and *Mathematica* commands for which the first few characters match those that you have typed.

Similarly, typing **Diode** and then selecting Make Template from the same submenu will append all the arguments required by **Diode** and will display
```
Diode[{node1, node2}, Rd, Cd, Rs, options]
```

You can then replace the prompts in the template with the actual values for your circuit, using the normal select-and-replace procedures for your machine. Make Template saves you from having to remember the arguments to functions or the attributes of a component and ensures that the arguments are in the correct order.

1.4.8 Options

Options are nonessential arguments that set up a context or ancillary information. For example, the **Plot** command has the option

> **PlotRange->{min, max}**

This tells *Mathematica* to display *y*-values between **min** and **max**, rather than to use *Mathematica*'s default method for deciding which part of the plotting range to display.

Toolbox

Rule symbol

-> (typed as a hyphen followed by a right angle bracket) is *Mathematica*'s rule symbol: **x->3** means "make **x** take the value of **3** only in the command or function in which the rule is being used." (If you wish to assign **x** the value **3** then use **x=3**.)

If you want information on the options for a command or function, then use the **Options** command to list the options:

> *In:*
>   **Options[Diode]**
>
> *Out:*
>   {Temperature -> 300}

To use an option, you merely enter its value in the function or component specification. For example, the diode component with and without the option of temperature value is as follows:

> **Diode[{1,14}, 10, 1 pF, 1, Temperature->270]**
>
> **Diode[{1,14}, 10, 1 pF, 1]**

If a function or component has no options, then **Options** returns an empty list:

> *In:*
>   **Options[I]**
>
> *Out:*
>   {}

1.4.9 Layout and Style

Mathematica is tolerant about layout style; you can lay out your input in a form that you find easy to read.

Toolbox

Comments

Any text between **(*** and ***)** will be treated as a comment; that is, *Mathematica* will ignore the intervening characters.

```
(*comment text*)
```

You can use the standard *Mathematica* comment syntax in any input intended for *Nodal*. Comments can be placed anywhere in an input cell. Comments at the beginning of a cell are useful for general description; you may find that on-the-line comments are useful for describing specific details or the use of a variable, for example. When using a Notebook interface to *Mathematica*, you can define whole cells to be text rather than input to *Mathematica*.

Within an evaluation cell, you can type in multiple commands before evaluating them (in series). Commands in series can be separated by semicolons. When commands are terminated by a semicolon, *Mathematica* suppresses the output resulting from those commands:

```
In:
  a=x+1

Out:
  1 + x


In:
  b=a^2

Out:
            2
  (1 + x)

In:
  a=x+1;
  b=a^2

Out:
            2
  (1 + x)
```

1.5 Error Messages

We all make mistakes: A finger will have landed on the wrong key or some other problem will arise, perhaps connected with syntax. Both *Mathematica* and *Nodal* have mechanisms that report error conditions or flag warnings.

For example, if you attempt to call a function and give incorrect arguments, you will receive an error message:

```
In:
  Sqrt[]

From In[1]:=
  Sqrt::argx:
      Sqrt called with 0
          arguments; 1 argument is expected.
```

The general style of error messages is the same: Messages indicate the function that reported the error, the message name (within that function), and explain what went wrong. If you want to learn more about error messages, we recommend that you read the Wolfram Research Technical Report *Mathematica* Warning Messages, which is supplied with *Mathematica*.

Mathematica checks your input carefully and detects misspellings. For example, if you mistype **Sqrt** (because your finger slipped onto the y key), then *Mathematica* will perform two actions. First, it will try to evaluate what you typed. Because *Mathematica* does not know a function with the name **Sqry**, it returns **Sqry** unevaluated. Second, *Mathematica* spots that **Sqry** is somewhat similar to **Sqrt** and issues a warning to you:

```
In:
  Sqry[1]

Out:
  Sqry[1]

From In:
  General::spell1:
      Possible spelling error: new symbol name "Sqry"
          is similar to existing symbol "Sqrt".
```

In general, you will probably find that such error messages and warnings are very helpful. If you want to stop *Mathematica* from issuing messages of a given type, you can do so with the **Off** function:

```
In:
  Off[General::spell1]
  Sqrg[1]

Out:
  Sqrg[1]
```

Mathematica will still return the undefined function **Sqrg** unevaluated. You can instruct *Mathematica* to resume reporting of messages by using **On**:

```
In:
  On[General::spell1]
  Sqre[2]

Out:
  Sqre[2]

From In:
  General::spell:
      Possible spelling error: new symbol name "Sqre"
       is similar to existing symbols
      {Sqrg, Sqrt}.
```

1.6 Summary

In this chapter, you have seen how

*   to get started with *Mathematica* and *Nodal*
*   to access on-line help
*   to describe and name components and circuits to *Nodal*
*   to use qualifiers to scale component values.

We also discussed the caveats that apply when nonlinear circuits are analyzed.

1.7 References

[Nolan53] Nolan, J., *Analytic Differentiation on a Digital Computer*, M.A. Thesis, Department of Mathematics, Massachusetts Institute of Technology, Cambridge, Massachusetts, 1953.

[Wolfram91] Wolfram, S., *Mathematica — A System for Doing Mathematics by Computer*, Addison-Wesley, Redwood City, California, 1991.

CHAPTER 2

DC Circuits

Although many electronic circuits are complex and contain perhaps thousands of components, they all consist of many small building blocks of which passive DC networks are the simplest. The design of even simple DC networks can be non-trivial, however, because their characteristics are affected by the resistance of circuitry to which they are connected, by nonlinear components, and by state-switching active components. For example, the basic potential divider is very sensitive to the load resistance that it feeds and, where AC signals are concerned, to unintentional (stray) circuit capacitances.

Using *Mathematica* and *Nodal*, you can compute parameters for simple DC circuits just as easily as for more complex, active circuits.

2.1 Voltage and Current in Resistive Circuits

Although it may seem almost too trivial to solve using a computer, *Mathematica* can assist in the application of Ohm's law. For example, if you know values for the voltage across (4 V) and current through (10 mA) a resistor, you can calculate the resistor's value in ohms by using *Mathematica* as an arithmetic calculator:

```
In:
  r=4/0.01

Out:
  400.
```

Because *Mathematica* is able to solve sets of equations, a more elegant approach is possible. You can define Ohm's law, numeric values of the current through and voltage across the resistor, and let *Mathematica* solve the equation for you.

Solving Equations

The `Solve` function requires two arguments: a list of equations to solve and a list of the variables in terms of which a solution is required. You can specify a list of variables to be eliminated as an optional third argument.

So that we can use again the symbol **r** symbolically, we first have to clear the numeric value it has been assigned:

```
In:
  Clear[r];
  Solve[{v==i r, v==4, i==0.01}, r]

Out:
  {{r -> 400.}}
```

Note that as arguments to `Solve`, each equation has a double = separating the left- and right-hand sides, like the equality test in the C programming language. You can also define, and then name, equations that you can pass as arguments to `Solve`:

```
In:
  ohmsLaw = (v==i r);
  Solve[{ohmsLaw, v==4, i==0.01}, r]

Out:
  {{r -> 400.}}
```

For a more substantial problem with a larger set of equations, consider a heating element made of a number, **n**, of resistors in series each of which is able to dissipate **w** watts, and a voltage supply of **vs** volts. For a required total output power, 15 W, how many resistors should be used and what value should they have (assuming that all the resistors have the same value, **r** Ω)? Given **vs**=30, **w**=1, **n** **w**=15 and, for equal load sharing among the resistors, **w**=(**vs/n**)^2/**r**, you can then use `Solve` to find **r** and **n**:

```
In:
  Solve[{n w==15,
         vs==30,
         w==((vs/n)^2)/r,
         w==1},
        {n,r}]
```

```
Out:
   {{r -> 4, n -> 15}}
```

If you leave **w** unspecified, the solution cannot be uniquely defined, so *Mathematica* gives the values for **r** and **n** in terms of **w**:

```
In:
   Solve[{n w == 15,
          vs == 30,
          w == ((vs/n)^2)/r},
         {n,r}]
```

```
Out:
                    15
   {{r -> 4 w, n -> -- }}
                    w
```

Note that symbolic answers give you a lot of information, even for trivial cases: **n** varies as the reciprocal of **w**, so, for small-valued **w**, you will need many resistors! Should you wish to choose a different **n** and **w**, you need not rework the analysis — you only have to reevaluate the symbolic answer. For large problems, reevaluation can often be substantially quicker than reanalysis.

In our next example, based on a simple potential divider, we show how you can specify which variables you want **Solve** to eliminate, and we introduce you to solving problems direct from a netlist, using *Nodal*.

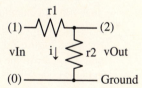

Figure 2.1   A simple resistive divider.

Given the resistive divider in Figure 2.1, a typical question might be: What is the output voltage (**vOut**) for a given input voltage (**vIn**)? We assume an open-circuit output — no load is connected.

The answer to this question requires the solution of two equations. From Ohm's law, the voltage across **r2** (that is, the output voltage) is the product of the current (**i**) and the resistance (**r2**) where the current through the circuit is the applied voltage (**vIn**) divided by the total resistance (**r1+r2**). You can use *Mathematica*'s **Solve** function to find a solution:

```
In:
   Solve[{i==vIn/(r1+r2), v2==r2 i}, v2]
```

Out:
```
    {{v2 -> i r2}}
```

Although this answer is correct, it is probably not what you intended: You asked for **v2** and you supplied an equation for it, so *Mathematica* used that information. You know **vIn**, **r1**, and **r2** but do not know **i**. You can eliminate **i** from the equation by specifying **i** as the third argument to **Solve** and so force *Mathematica* to solve the two equations in the manner you intended:

In:
```
  Solve[{i==vIn/(r1+r2), v2==r2 i},
        v2, i] //Simplify
```

Out:
```
              r2 vIn
    {{v2 -> -------}}
              r1 + r2
```

Until now, we have asked *Mathematica* to solve equations culled from our knowledge of circuit analysis. You proceed differently when you use *Nodal*. *Nodal* requires only that you provide a netlist description of the circuit and what particular characteristic you wish to determine. (In Chapter 4, we discuss how *Nodal* uses netlists to model circuit behavior.) For example, how do you use *Nodal* to determine the relative output voltage of the circuit in Figure 2.1? First, if you have not loaded *Nodal*, you must do so by issuing the command **Needs**. Then, you define a netlist as the argument to the **NodalNetwork** function. So that you can calculate the output voltage, you must declare a voltage source somewhere in the circuit — connected between node 1 and ground (node 0) as shown in Figure 2.1. If you assign a name to the circuit (say, **rDivider**), later you can refer to the circuit by its name, rather than having to respecify the network:

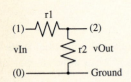

In:
```
  Needs["Nodal`","Nodal2.m"];
  rDivider=NodalNetwork[
             VoltageSource[{1,0},vIn],
             Resistor[{1,2},r1],
             Resistor[{2,0},r2]]
```

Second, to calculate the ratio of the voltage at node 2 to that at node 1, you use the *Nodal* function **NodalAnalyze** with the circuit that is to be analyzed, called **rDivider**, as the first argument; the second argument defines what you want the result of the analysis to be: the ratio of the voltage at node 2 to the voltage at node 1, **V2/V1**:

In:
```
  NodalAnalyze[rDivider, Result->V2/V1]
```

```
Out:
            1
    -(---------------)
           1     1
    r1 (-(--) - --)
          r1     r2
```

    Mathematica sometimes leaves answers in a format that is not minimized. You can reduce this answer to a simpler and more recognizable form using *Mathematica*'s **Simplify** command.

Toolbox

Recalling the Last Answer

In *Mathematica*, you can use the percent (**%**) symbol to refer to the last answer; using **%** saves you from having to name all results and is convenient when you wish to refer to a result only once, perhaps immediately after that result has been calculated by the previous operation. To recall the next-to-last answer, you use **%%**.

Simplifying Expressions

You may often find that the **Simplify[**expression**]** function clarifies a complicated *expression*.

Other useful *Mathematica* commands that alter the form of an answer are **Expand**, **ExpandAll**, **Factor**, **Together**, **Apart**, and **Cancel**.

```
In:
    Simplify[%]
Out:
          r2
        -------
        r1 + r2
```

    There are some important points to note about using **NodalAnalyze**. If you want to determine a number of, say, voltages in your circuit, you must place them in a list: **Result->{V1, V3, V5, I8, V5/V1}**. The function **NodalAnalyze** is one of *Nodal*'s main functions and allows you to specify a wide range of circuit characteristics for the **Result** option; we give more information on **NodalAnalyze** in the Appendix.

If you want to use answers from **NodalAnalyze** in other calculations, **NodalAnalyze** can return answers in the form of a nested list. You can display the complete form of the answer from any *Mathematica* function, including those supplied in *Nodal*, by using the function **Normal**. For example, you can see that **NodalAnalyze** has returned your answer as the second element of a two-member list, which is, in turn, the first member of a single-element list:

```
In:
  Normal[%]

Out:
            r2
  {{1,  -------}}
          r1 + r2
```

Wrapping the answer in a list within a list may seem to be a needless complication, but there is a reason, of course. In AC circuit analysis, you can use **NodalAnalyze** to calculate the voltages, say, at several different nodes in a circuit and repeat the calculation at a number of different frequencies. The answer returned from **NodalAnalyze** is therefore a two-dimensional array and is represented in *Mathematica* as a list of lists. In your analysis of **rDivider**, you asked that one voltage be calculated. By default, **NodalAnalyze** has analyzed the circuit using the symbolic value **f** for frequency; **f** has not appeared in the answer, because the answer is frequency independent. Hence, your analysis of **rDivider** has resulted in a list containing one member, because only one frequency value was used, and that member has two elements: a number, identifying the member as the first member, and the single answer, the voltage at node 2.

2.1.1 Mesh Analysis

Networks may contain multiple voltage sources and loops. You can use *Mathematica* to solve for the various circuit parameters in these more complex circuits in any of three ways. First, you can apply Kirchhoff's laws, derive the mesh equations manually, and then use **Solve**. Second, you can write the mesh equations explicitly in matrix form and then use *Mathematica* to perform the necessary matrix algebra. Third, using *Nodal*, you can proceed directly from the netlist to your result.

Kirchhoff formulated two fundamental laws of circuit analysis. These laws describe how voltages and currents behave in a network [Adby80]. (In hindsight these are actually abstractions of Maxwell's field equations, but Kirchhoff's equations came first.) First, there is Kirchhoff's voltage law which states that the sum of the voltages around any loop in a circuit must be equal to zero and is derived from Maxwell's equation, $\int Edl + d\phi/dt = 0$, where E is the electric field strength, dl is the line segment length, and ϕ is the magnetic flux density. Second, there is Kirchhoff's current law which states the sum of the currents into any circuit node must be equal to zero and is derived from Maxwell's equation $\int J\,ds + d\rho/dt = 0$, where J is the current density, ds is the area with current flowing through it, and ρ is the charge density. Application of either of these two laws allows you to analyze a circuit completely. Analysis based on voltages around a loop is called *mesh analysis*, and analysis based on currents into a node is called *nodal analysis*. In either case, you can write a set of equations describing each loop in the circuit or each node in the circuit. These equations are often written in a matrix: The loop analysis forms an impedance matrix; the nodal analysis forms an admittance matrix. We prefer the admittance matrix description because you can determine the matrix coefficients by inspection from the component values and their nodal end points; the impedance matrix description is harder to program on the computer. If you are interested, you might want to pursue the connections between graph theory and analysis via meshes or nodes [Vlach83].

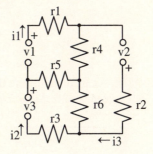

Figure 2.2   Mesh circuit.

Using mesh analysis, you can describe the circuit in Figure 2.2 by the following three equations:

```
(r1+r4+r5)i1 - r4 i2 - r5 i3  = v1
-r4 i1 + (r2+r4+r6)i2 - r6 i3 = v2
-r5 i1 - r6 i2 + (r3+r5+r6)i3 = v3
```

You can find the solution (for the currents **i1**, **i2**, and **i3**) by using **Solve**. We have omitted *Mathematica*'s full symbolic answer, which is rather long and not very interesting, by terminating our input with a semicolon:

```
In:
    Solve[{i1 (r1+r4+r5) - i2 r4 - i3 r5 == v1,
           -i1 r4 + i2 (r2+r4+r6) - i3 r6 ==v2,
           -i1 r5 -i2 r6 + i3 (r3+r5+r6) == v3},
          {i1,i2,i3}];
```

If you require a numerical result, then you can specify the values of components before, in, or after the **Solve** command. For simplicity, the following example uses a value of 1 Ω for **r1**, 2 Ω for **r2**, . . ., 6 Ω for **r6** and, in a similar manner, 1 V for **v1**, . . ., 3 V for **v3**.

```
In:
  r1=1;
  r2=2;
  r3=3;
  r4=4;
  r5=5;
  r6=6;
  v1=1;
  v2=2;
  v3=3;
  Solve[{i1 (r1+r4+r5) - i2 r4 - i3 r5 == v1,
         -i1 r4 + i2 (r2+r4+r6) - i3 r6 ==v2,
         -i1 r5 -i2 r6 + i3 (r3+r5+r6) == v3},
         {i1,i2,i3}]
Out:
  {{i1 -> 1, i2 -> 1, i3 -> 1}}
```

As an alternative to defining the variable values before you invoke **Solve**, you can use *Mathematica*'s replacement operator.

Toolbox

Prefix, Infix, and Postfix Form of Functions

As an alternative to the standard format for operator and argument, operators can be used in prefix, infix, and postfix forms:

standard: **FunctionName** [*argument*]

prefix: **FunctionName** @ *argument*

infix: *arg1* ~**FunctionName**~ *arg2*

postfix: *argument* **//** **FunctionName**

Replacement

The infix form of the **ReplaceAll** function, **/.**, is used to replace symbols in an expression on its left by values described by rules (like *symbol->value*) on its right.

expression **/.** *rules*

Taking Parts of Lists

Given a list **L** that contains **n** members, the *i*th member can be identified by **L[[*i*]]**, where 1≤*i*≤**n**.

Numerical Operator

N[*argument***]** forces *Mathematica* to give the floating-point form of its argument. The postfix form is **//N**.

N[*argument***,** *sigFigs***]** will format *argument* as a floating-point number with *sigFigs* number of significant figures.

In the next example, you establish rules for the values of **i1**, **i2**, **i3** using **Solve**. Then, after the semicolon, those rules (which have become the last answer and so referred to by the **%** symbol) have their symbols replaced by the values assigned in the rules following the replacement operator (**/.**). The answers are real numbers, identifiable by the trailing decimal point (integer answers would not be followed by a decimal point) because the enumerate operator, **N**, has been applied in postfix form (**//N**):

```
In:
   Solve[{i1 (r1+r4+r5) - i2 r4 - i3 r5 == v1,
          -i1 r4 + i2 (r2+r4+r6) - i3 r6 ==v2,
          -i1 r5 -i2 r6 + i3 (r3+r5+r6) == v3},
          {i1,i2,i3}];

   % /. {r1->1, r2->2, r3->3, r4->4, r5->5,
               r6->6, v1->1, v2->2, v3->3} //N

Out:
   {{i1 -> 1., i2 -> 1., i3 -> 1.}}
```

Mathematica, by default, often displays numbers with an implied accuracy that may be inappropriate. You can control the precision with which numerical output is displayed when using **N** by specifying the required number of significant digits as the second argument to **N**:

```
In:
   N[2.266 Pi]
Out:
   7.11885
```

In:
  N[2.266 Pi, 2]

Out:
  7.1

In:
  N[2.266 Pi, 3]

Out:
  7.12

You can also solve the problem by writing the equations explicitly in the form of a matrix equation and then asking *Mathematica* to invert the matrix containing the resistor values:

$$\begin{bmatrix} (r1 + r4 + r5) & -r4 & -r5 \\ -r4 & (r2 + r4 + r6) & -r6 \\ -r5 & -r6 & (r3 + r5 + r6) \end{bmatrix} \begin{bmatrix} i1 \\ i2 \\ i3 \end{bmatrix} = \begin{bmatrix} v1 \\ v2 \\ v3 \end{bmatrix}$$

Representing each matrix by a single-character name, you can write

R I = V

You can then find the currents by premultiplying both sides of the equation by the inverse of **R**:

$$\mathbf{R}^{-1} \ \mathbf{R} \ \mathbf{I} = \mathbf{R}^{-1} \ \mathbf{V}$$

so

$$\mathbf{I} = \mathbf{R}^{-1} \ \mathbf{V}$$

You can solve matrix equations like these using *Mathematica*'s functions for inverting, transposing, and multiplying matrices.

Toolbox

Matrix Multiplication

Matrices can be multiplied together using the **Dot** [*arg1, arg2*] operator, which can be contracted to its infix form . to mimic normal mathematical notation.

arg1 . arg2

Matrix Inversion and Transposition

Inverse[*matrix*] and **Transpose**[*matrix*] compute the inverse and transpose of *matrix*.

MatrixForm[*matrix*] prints *matrix* in mathematical notation:

```
In:
    myMatrix={{a1,b1,c1},
             {d,e,f}};
    MatrixForm[myMatrix]
```
```
Out:
    a1    b1    c1
    d     e     f
```
```
In:
    myVector={{a2},{b2},{c2}};
    MatrixForm[myVector]
```
```
Out:
    a2
    b2
    c2
```

In this example, **R** is the resistor matrix. **v** is initially defined (arbitrarily) as a 1×3 (that is, one row, three columns) matrix, **{{1,2,3}}**, but is required as a 3×1 matrix, **{{1},{2},{3}}**, so that it can be premultiplied by the 3×3 matrix **R**. Hence you have to use **Transpose**[] to alter the dimensions of **v**. The 3×1 current matrix containing values for the currents i1, i2 and i3 is found by premultiplying **v** by the inverse of **R**, named **iR**:

```
In:
    rArray={{10,-4,-5},
            {-4,12,-6},
            {-5,-6,14}};
    vArray=Transpose[{{1,2,3}}];
    invRArray=Inverse[rArray];
    invRArray.vArray
```
```
Out:
    {{1}, {1}, {1}}
```

Because you now know the current, resistor, and voltage values around the circuit, you can compute, say, the voltage at any node measured with reference to some (arbitrarily) chosen node.

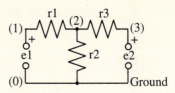

Figure 2.3   A nodal network.

You can also perform mesh analysis calculations using *Nodal*. For example, in Figure 2.3, you can calculate the voltage at node 2 by defining the nodal network to be used and then requesting the result to be the voltage at node 2, **v2**, when the circuit is analyzed. We have named the voltages produced by the voltage sources **e1** and **e2** to differentiate them from the variable name **v2**, which *Nodal* interprets as the voltage at node 2, and cleared any assignments to **r1**, **r2**, or **r3**.

In:
```
Clear[r1,r2,r3];
test=NodalNetwork[VoltageSource[{1,0},e1],
                  Resistor[{1,2},r1],
                  Resistor[{2,0},r2],
                  Resistor[{2,3},r3],
                  VoltageSource[{3,0},e2]];
result=NodalAnalyze[test,
                    Result->V2]//Simplify
```

Out:
```
r2 (e2 r1 + e1 r3)
--------------------
r1 r2 + r1 r3 + r2 r3
```

You can also determine the current through a node using **Result->Ij**, where **j** is the node number *and the node is adjacent to a grounded voltage source*. Note that while **Ij** is used to represent the current at node **j**, the single character **I** represents $\sqrt{-1}$.

In:
```
div=NodalNetwork[VoltageSource[{1,0},10],
                 Resistor[{1,2},10],
                 Resistor[{2,0},10],
                 Resistor[{2,3},5],
                 Resistor[{3,4},5],
                 VoltageSource[{4,0},0]];
NodalAnalyze[div,Result->{I1,I2,I4,V3,V3/5}]
```

Out:

Step	I1	I2	I4	V3	$\dfrac{V3}{5}$
1.	0.667	0	-0.333	1.67	0.333

If you use a current source of zero output current instead of a voltage source then, because current sources have infinite impedance, the current source effectively blocks off that branch of the network:

In:
```
div=NodalNetwork[VoltageSource[{1,0},10],
                 Resistor[{1,2},10],
                 Resistor[{2,0},10],
                 Resistor[{2,3},5],
                 Resistor[{3,4},5],
                 CurrentSource[{4,0},0]];
NodalAnalyze[div,Result->{I1,I2,I4,V3,V3/5}]
```

Out:

Step	I1	I2	I4	V3	$\dfrac{V3}{5}$
1.	0.5	0	0	5.	1.

2.1.2 Multistate DC Circuits

You can use *Mathematica* to calculate component values in multistate DC circuits. For example, if you design a Schmitt trigger, you will model its circuit with a multistate DC network. A Schmitt trigger is often used to monitor an analog signal: When the signal drops below a threshold, the circuit switches on — and switches off when the signal rises above the threshold. To prevent noise in the analog signal from causing rapid switching (perhaps at frequencies up to several megahertz), hysteresis is applied to the threshold. When the trigger switches on, the threshold level is increased by a small amount and vice versa when the circuit switches off.

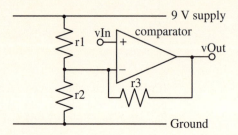

Figure 2.4   Voltage comparator.

A voltage comparator (see Figure 2.4) is commonly used to implement a Schmitt trigger. The threshold voltage is supplied to the comparator's inverting input. In this example, when the comparator's output is low, it is at 1 V; when it is on, it is 1 V less than the supply voltage.

The midpoint of the Schmitt action (which might be referred to as the nominal threshold) is the voltage at the inverting input when v_{out} is at half the supply voltage. The actual thresholds will be at an offset voltage below (the switch-on value) and above (the switch-off value) the nominal threshold; the adjustment of the threshold is achieved by feedback through **r3**. The design task is to calculate the correct values of the three resistors in Figure 2.4. To calculate resistor values you can either apply Kirchhoff's voltage law to the circuit or use *Nodal*.

Applying Kirchhoff's voltage law to the simple network in the three situations (inverting input at the nominal, low, and high threshold values) gives you nine equations: The currents through the arms of the circuit are different in each of the three situations (we identify them by the names **i1**, **i1a**, and **i1b**). *Mathematica* can solve the nine equations to give you the values of the resistors that would be required to implement the circuit:

```
In:
  Solve[{9==i1 r1+(i1+i2)r2,
         6==(i1+i2)r2,
         4.5==i2 r3+(i1+i2)r2,
         9==i1a r1+(i1a+i2a)r2,
         5.9==(i1a+i2a)r2,
         1==i2a r3+(i1a+i2a)r2,
         9==i1b r1+(i1b+i2b)r2,
         6.1==(i1b+i2b)r2,
         8==i2b r3+(i1b+i2b)r2},
         {r1,r2,r3},{i1,i1a,i1b,i2,i2a,i2b}]
```

```
Out:
   {{r3 -> 11.1667 r2, r1 -> 0.489051 r2}}
```

The answer shows that the values of the resistors are not fixed in an absolute sense — only their relative values are important: You are free to choose a value for **r2** that will then prescribe the value of **r1** and **r2**. The value of **r2** will define the typical current that will flow through the circuit. Your choice of current may be governed by such considerations as power available, the input and output current capability of the comparator chip, and the required extent of immunity to ambient noise.

You can also use *Mathematica* to verify the voltages at the inverting input that will result from using the circuit values already solved by *Mathematica* (assuming **r2**=10 kΩ):

```
In:
   Solve[{9==4890 i1 + (i1+i2) 10000,
          1==(i1+i2) 10000 + i2 111700},
         {i1,i2}] //N

Out:
   {{i1 -> 0.000633895, i2 -> -0.0000438698}}


In:
   10000 (i1+i2) /. %

Out:
   {5.90025}


In:
   Solve[{9==4890 i1 + (i1+i2) 10000,
          8==(i1+i2) 10000 + i2 111700},
         {i1,i2}] //N

Out:
   {{i1 -> 0.00059301, i2 -> 0.0000170082}}


In:
   10000 (i1+i2) /. %

Out:
   {6.10018}
```

Alternatively, to determine the values for the resistors we can ask *Nodal* to calculate the symbolic equation for the voltage at node 2 and then use **Solve** to find the values for **r1**, **r2**, and **r3** that give the voltages we require when **vOut** is set to its required values.

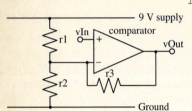

In:
```
mySchmitt=NodalNetwork[VoltageSource[{1,0},9],
                       Resistor[{1,2},r1],
                       Resistor[{2,0},r2],
                       Resistor[{2,3},r3],
                       VoltageSource[{3,0},vOut]];
vNode2=(NodalAnalyze[mySchmitt,Result->V2]//
                       Normal)[[1,2]]
```

Out:
```
        9                   vOut
---------------- + ----------------
 1    1    1       1    1    1
r1 (-- + -- + --)  (-- + -- + --) r3
   r1   r2   r3     r1   r2   r3
```

In:
```
rValueRules=Solve[{(vNode2 /. vOut->1)==5.9,
          (vNode2 /. vOut->8)==6.1}, {r1,r2,r3}]
```

Out:
```
{{r1 -> 0.0437956 r3, r2 -> 0.0895522 r3}}
```

As before, you have to choose a value for **r3**. Let **r3** be 10 Ω. You can then check
the voltage on node 2, when **vOut** is in its low and high state:

In:
```
vNode2 //.
    Flatten[{r3->10,
            rValueRules[[1]]}] /. vOut->{1,8}
```

Out:
```
{5.9, 6.1}
```

2.2 Power-Conversion Utilities

Conversions between power dissipated in, current through, and voltage across a
specified impedance are common calculations in both DC and AC circuits. Using
Mathematica, you can easily construct time-saving functions to perform these
conversions.

Toolbox

Introducing User-Defined Functions

The ability to define your own functions is very useful. For simple tasks, functions act like miniprograms and allow you to avoid reworking calculations. Once a function is defined, you can use it immediately.

User-defined functions have the general form *functionName* [*x_*] : = *functionBody*. For specific types of argument, *x_* becomes *x_type*, where *type* is one of *Mathematica*'s recognized types.

The definition of functions is an important topic and we give more information in the Appendix; we also recommend that you read more about it in Stephen Wolfram's book *Mathematica — A System for Doing Mathematics by Computer* [Wolfram91].

Here is an example of a function that converts voltage across a resistance to a power:

```
In:
  VRtoWatts[voltage_,resistance_]:=
                          voltage^2/resistance


In:
  VRtoWatts[15,100]
Out:
            9
            -
            4


In:
  VRtoWatts[volts,2000]
Out:
          2
      volts
      ------
       2000
```

Nodal provides several utilities to convert between power and volts or amperes (through a resistance you specify). All the utilities in *Nodal* are named in a straightforward manner; we list them in the Appendix. For example, you can calculate the power dissipated by a 100 Ω resistor dropping 15 V by using the `Watt` function:

```
In:
  Watt[Volt[15], Zo->100]
```

N

```
Out:
            9
            -
            4
```

Because the current and impedance also uniquely define the power, the same utility can be used with current as its first argument:

```
In:
  Watt[Ampere[3/20], Zo->100]
```

N

```
Out:
            9
            -
            4
```

Mathematica will try to work with the simplest and most precise form of a number — which can mean that the answer is left as a vulgar fraction. If such a form is inconvenient, then you can obtain a floating-point version of the answer by using the enumerate function **N** in its prefix or postfix format:

```
In:
  Watt[Volt[15], Zo->100] //N
```

N

```
Out:
  2.25
```

```
In:
  Watt[Volt[15], Zo->100] //N
```

N

```
Out:
  2.25
```

The current passing through or the voltage across the resistor can be found if you know the power dissipated. You can also work backward, from power into an impedance, to volts.

```
In:
  Ampere[Watt[9/4], Zo->100]
```

N

```
Out:
            3
            --
            20
```

```
In:
    Volt[Watt[9/4], Zo->100]
```

N

```
Out:
    15
```

Nodal has utilities for calculating dB from power or voltage ratios. A factor of four in power is equivalent to a 6 dB increase; a factor of 4 in voltage or current is equivalent to a 12 dB increase:

```
In:
    DBP[4]
```

N

```
Out:
    6.0206
```

```
In:
    DB[4]
```

N

```
Out:
    12.0412
```

You can also convert an arbitrary power level, referenced to some standard level, to dB. The *Nodal* utility **DBRef[]** assumes a standard reference level of 1 mW. A 1 W power dissipation is 30 dB up on that level.

```
In:
    DBRef[Watt[1]]
```

N

```
Out:
    30.
```

You can set both the reference level and the impedance into which the power is fed to values specific for your application by using the options on the **DBRef** command. For example, if your reference power level is 100 mW into 68 Ω, you can calculate the relative power (in dB) that will result from applying a 5 V signal to the same impedance:

```
In:
    DBRef[Volt[5],
          Reference->100 Milliwatt, Zo->68]
```

N

```
Out:
    5.65431
```

Nodal also provides an inverse utility that will convert a power comparison in dB into a relative voltage level. What is the relative voltage across the system that corresponds to a −5 dB power level, when referenced to your defined 0 dB level?

```
In:
    ArcDB[-5]
```

N

```
Out:
    0.562341
```

Lowering the applied voltage to 0.56 times its present value will result in a 5 dB reduction in power. A 6 dB reduction in power is synonymous with a reduction of the power to ~0.25 times its present value:

```
In:
    ArcDBP[-6]
```

N

```
Out:
    0.251189
```

You can also use utilities to answer design questions. For example, what voltage will result in a 9 dB power level in a system where the 0 dB level is equated to 100 mW dissipation in 68 Ω?

```
In:
    Solve[DBRef[Volt[x],
              Reference->100 Milliwatt,
              Zo->68]==9,x]
```

N

```
Out:
                     34              (-1)
    {{x -> -(Sqrt[--] Sqrt[DBP    [9]])},
                     5


                    34              (-1)
    {x -> Sqrt[--] Sqrt[DBP    [9]]}}
                    5
```

Because power is proportional to the square of the applied voltage, either a positive or a negative voltage could result in a dissipation of 9 dB. *Mathematica* has given you both answers. The symbol **DBP**$^{(-1)}$ refers to the inverse **DBP** function; *Mathematica* does not know automatically either that there is an inverse function or its name, but *Mathematica* does tell you that the inverse of **DBP** is required. *Nodal* provides an inverse: **ArcDBP**:

```
In:
    Sqrt[(34/5) ArcDBP[9]]
```

N

```
Out:
    7.34944
```

2.3 Attenuator Design

The design of T- and π-attenuators, or pads, is merely the solution of Kirchhoff's laws applied around the various loops in the attenuator circuit. Common symmetrical T- and π-pads are shown in Figure 2.5.

You can write several simple equations for the currents and voltages in the circuit and then use *Mathematica* to solve for **r1** and **r2**, by eliminating **i1**, **i2**, **i3**, and **i4**, given the attenuation factor (a factor of 10, a 20 dB loss, is explicitly given last in the list of equations) and the input and output impedances, **z**:

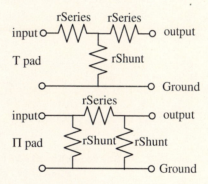

Figure 2.5   T- and π-pad networks.

```
In:
  Solve[{i1 == i2+i3+i4,
         vIn == i2 r2,
         i2 r2 == r1(i3+i4)+i3 r2,
         i3 r2 == i4 z,
         vOut == i4 z,
         vIn == i1 z,
         vOut == vIn/10},
        {r1,r2},{i1,i2,i3,i4}]

Out:
            99 z           11 z
  {{r1 -> ----,  r2 -> ----}}
            20             9
```

If you want to design a pad for other attenuation factors, you merely have to substitute in the new value of attenuation.

Nodal has a utility that allows you to calculate the resistor values in T- and π-pads. When using the pad-designing utility, you must specify the required matching impedance and whether you require a T or π geometry. For example, to compute the series and shunt resistors in a 20 dB T attenuator in a 100 Ω system,

In:
```
Pad[20, Zo->100, Type->T]
```

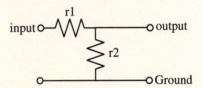

Out:
```
{Rseries -> 81.8182, Rshunt -> 20.202}
```

Mathematica can help you with the design of other attenuator types. The L-attenuator, shown in Figure 2.6 is possible only for particular values of source, **rS**, and load, **rL**, impedance. First, you can use **Solve** to calculate the values of **r1** and **r2** as a function of **rS** and **rL**, given the constraint that the impedance looking into a port must equal the impedance that will be connected to that port:

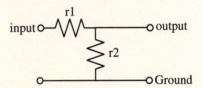

Figure 2.6   L-attenuator.

In:
```
Solve[{rS==r1+1/(1/r2+1/rL),
       rL==1/(1/r2+1/(r1+rS))}, {r1,r2}]
```

Out:

```
                                   2
                                 rL
      {{r1 -> -rL + rS + --------------------,
                             rL Sqrt[rS]
                       rL - --------------
                            Sqrt[-rL + rS]

                 rL Sqrt[rS]
       r2 -> -(--------------)},
               Sqrt[-rL + rS]


                                   2
                                 rL
       {r1 -> -rL + rS + --------------------,
                             rL Sqrt[rS]
                       rL + --------------
                            Sqrt[-rL + rS]

               rL Sqrt[rS]
       r2 -> --------------}}
             Sqrt[-rL + rS]
```

You can see a constraint on the design. For **r2** to be positive, real, and finite, **rS** must be greater than **rL**. If you know numeric values for **rS** and **rL**, you can find **r1** and **r2** and the numeric attenuation factor (–10 dB). We have to use **Flatten** because the replacement operator (**/.**) expects a single list of rules and we have two lists:

```
In:
  resistorValues=% /. {rS->30,rL->10} //N
Out:
  {{r1 -> -24.4949, r2 -> -12.2474},
   {r1 -> 24.4949, r2 -> 12.2474}}
```

```
In:
  1/(1/r2+1/rL) /
      (rS+r1+(1/r2+1/rL)) /.
    Flatten[{resistorValues[[2]],
             {rS->30,rL->10}}]
Out:
  0.100685
```

2.4 Nonlinear DC Circuits

The concept of linearity in circuit operation is both important and simple: In a linear circuit, the output voltage or current, y, is proportional to an input or excitation voltage or current, x — if $y = f(x)$ then $cy = f(cx)$. For example, if you double the voltage across an ideal resistor then the current passed will also double.

However, many electrical and electronic devices (such as light bulbs and diodes) exhibit nonlinear behavior. If you are to be able to model and analyze circuits containing nonlinear devices, you need to characterize the devices by determining an expression for the current passed by the device as a function of the applied voltage.

Mathematica can help you with the task of fitting, or deriving, a formula from experimental data. You can then use the derived function to model circuit behavior. Electronics is a very practical subject, and *Mathematica* provides a rich tool set with which to tackle real-world engineering problems. It is equally at home with your experiment's data as with a complicated engineering mathematical model.

2.4.1 Device Characterization Using Experimental Data

Data from experiments are stored most simply in ASCII files. For example, using the circuit shown in Figure 2.7, the current through a silicon diode was measured as a function of the voltage across the diode.

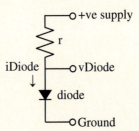

Figure 2.7   Circuit used for measuring diode's data.

The following table of values was obtained and typed into a simple ASCII file with line breaks at the end of each line.

```
0.112 0.536
0.087 0.524
0.158 0.550
0.222 0.565
0.296 0.576
0.348 0.585
0.742 0.620
0.957 0.630
1.420 0.650
2.350 0.673
4.820 0.708
8.010 0.734
```

For some reason, the current (in milliamperes) was placed first in each line, with the second value being the voltage across the diode (in Volts). The file, recorded on an Apple Macintosh, is called "Si diode iv data"; other computers might not accept this form of file name (with embedded spaces), so, prior to using the file-handling facilities in *Mathematica*, you must consider how the computer system that you are using normally names files.

Reading Data from ASCII Files

Use **ReadList** [*"filename", type*] to read in data of a given type from the file called *filename*. More information about reading data from files is given in the Appendix.

To read in the current and voltage data as a flat-structured list of real numbers, you specify the type of data to be **Number**:

```
In:
    ReadList["Si diode iv data",
             Number]
```

```
Out:
    {0.112, 0.536, 0.087, 0.524, 0.158, 0.55, 0.222, 0.565,
    0.296, 0.576, 0.348, 0.585, 0.742, 0.62, 0.957, 0.63, 1.42,
    0.65, 2.35, 0.673, 4.82, 0.708, 8.01, 0.734}
```

When *Mathematica* reads in data from a file, you can specify other types of objects. Valid types include **Byte**, **Character**, **String**, **Real**, and **Number**. If you specify **Expression**, *Mathematica* will read in a complete *Mathematica* expression. **Real** expects an approximate number in FORTRAN-like E format and **Number** expects either an exact number (that is, an integer) *or* an approximate number (equivalent to **Real**).

Of course, the data in the file "Si diode iv data" are actually pairs of numbers and it is better if this structure is maintained. You can read in numbers by pairs (or by analogy, in any other form of grouping) by indicating the form of the grouping in the **ReadList** function:

```
In:
    diodeIV=ReadList["Si diode iv data",
                     {Number,Number}]
```

```
Out:
    {{0.112, 0.536}, {0.087, 0.524}, {0.158, 0.55}, {0.222,
    0.565}, {0.296, 0.576}, {0.348, 0.585}, {0.742, 0.62},
    {0.957, 0.63}, {1.42, 0.65}, {2.35, 0.673}, {4.82, 0.708},
    {8.01, 0.734}}
```

After you defined the structure to be number pairs — **{Number, Number}** — *Mathematica* read in the data as pairs and placed them into a list of pairs. We gave the list a name, **diodeIV**, to make referring to it easier. Note that the data have not been ordered. However, the data are still in an inconvenient format for plotting, because each pair is {current(voltage), voltage} and *Mathematica* expects plotting points to be paired as {x, y}. To correct the in-pair order, you can use *Mathematica*'s list-manipulating functions. We would normally merge the next three operations into one line, but we have left them as individual actions so that you can see how each affects the contents of the lists:

In:
```
Transpose[diodeIV]
```

Out:
```
{{0.112, 0.087, 0.158, 0.222, 0.296, 0.348, 0.742, 0.957,
1.42, 2.35, 4.82, 8.01}, {0.536, 0.524, 0.55, 0.565, 0.576,
0.585, 0.62, 0.63, 0.65, 0.673, 0.708, 0.734}}
```

In:
```
Reverse[%]
```

Out:
```
{{0.536, 0.524, 0.55, 0.565, 0.576, 0.585, 0.62, 0.63,
0.65, 0.673, 0.708, 0.734}, {0.112, 0.087, 0.158, 0.222,
0.296, 0.348, 0.742, 0.957, 1.42, 2.35, 4.82, 8.01}}
```

In:
```
dIV=Transpose[%]
```

Out:
```
{0.536, 0.112}, {0.524, 0.087}, {0.55, 0.158}, {0.565,
0.222}, {0.576, 0.296}, {0.585, 0.348}, {0.62, 0.742},
{0.63, 0.957}, {0.65, 1.42}, {0.673, 2.35}, {0.708, 4.82},
{0.734, 8.01}}
```

You can now plot the data for inspection (details of graphics functions are given in the Appendix):

In:
```
dataplot=ListPlot[dIV,
  AxesLabel->{"Volts","mA"},
  AxesOrigin->{0.52,0.0},
  PlotLabel->"Diode current vs voltage"]
```

Out:

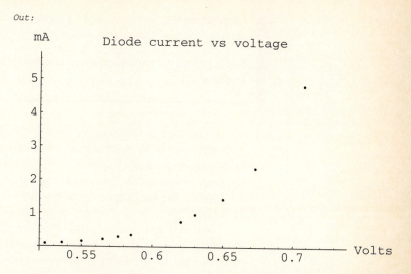

We left to *Mathematica* the task of choosing the range over which to plot, but you can control the range plotted, in either axis, by using the **PlotRange** option in the **ListPlot** command. For example, if you specify **PlotRange->** **{{0.4, 0.8}, {0, 10}}**, the range of each axis is defined: The *x*-axis would show from 0.4 V to 0.8 V and the *y*-axis from 0 mA to 10 mA. The **AxesOrigin** option allows you to specify the coordinates at which the *x*- and *y*-axes intersect.

Once *Mathematica* has read in the data, your next step in characterizing the current–voltage relationship is to fit an analytical function to the data. The current through a diode is given (to a first approximation [Delaney69]) by the equation **i = io Exp[av]** where **io** and **a** are constants and **v** is the voltage across the diode. To express this equation as a linear equation in **v**, you take logarithms of both sides:

Log$_e$(i) = Log$_e$(io) + av

The term **Log$_e$(io)** will be a constant.

Toolbox

Fitting Data to Functions

Fit [*data, function, var*] will fit a function to data as a dependent of the variable *var*.

Logarithms

Log [x] gives the natural logarithm of x. For logarithms to other bases, use
Log [$base, x$].

To use the **Fit** function, you need to take the (natural) logarithms of the currents measured in the preceding experiment. The data to which the fit will be made must be a list of $\{x,y\}$ pairs — in this case, $\{v, \text{Log}_e(i)\}$. In the following sequence of commands, we demonstrate some of *Mathematica*'s list-manipulating abilities; the use of anonymous functions (discussed in Chapter 4) would be far more efficient.

First, so that you can take their logarithms, you need to group into one list all the current values, rather than having them distributed in each pair. The transpose of the N-long list of number pairs that was read in will produce a $2 \times N$ matrix, the first, or top, row of which will contain the current samples. We can then apply the **Log** function to the list of current samples and then use **Transpose** again to convert the pair of lists back to a list of pairs.

```
In:
  {voltages,currents}=Transpose[dIV]

Out:
  {{0.536, 0.524, 0.55, 0.565, 0.576, 0.585, 0.62,
    0.63, 0.65, 0.673, 0.708, 0.734},
   {0.112, 0.087, 0.158, 0.222, 0.296, 0.348,
  0.742, 0.957, 1.42, 2.35, 4.82, 8.01}}
```

```
In:
  Transpose[{voltages,Log[currents]}]

Out:
  {{0.536, -2.18926}, {0.524, -2.44185},
   {0.55, -1.84516}, {0.565, -1.50508},
   {0.576, -1.2174}, {0.585, -1.05555},
   {0.62, -0.298406}, {0.63, -0.0439519},
   {0.65, 0.350657}, {0.673, 0.854415},
   {0.708, 1.57277}, {0.734, 2.08069}}
```

The output of the last **Transpose** command now contains a list of {voltage, Log_e(current)} pairs. Now that the data are in a form that can be used by **Fit**, you can obtain an analytical expression for the data:

```
In:
  eqn=Fit[%,{1,v},v]
```

Out:
>   -13.7368 + 21.6437 v

That is, $i(V) = e^{-13.7} e^{21V}$ mA.

Lists are important and often-used data structures in *Mathematica*. If *Mathematica* is new to you, then you may find that it takes some time before you feel comfortable using lists — which is why we have separated many of the list operations so far. Once you feel confident about handling lists, you can perform all the listhandling required for plotting and fitting the diode data in a few lines.

Just to demonstrate more efficient use of list handling, we will plot the measured diode data in amperes on both linear and logarithmic scales. By dividing the **currents** list by 1000, we obtain a function that gives the natural logarithm of the current in amperes (remember that the data were in mA) as a function of the applied voltage:

In:
```
{currents,voltages}=Transpose[diodeIV];
ListPlot[Transpose[{voltages,currents}],
        AxesLabel->{"volts","mA"}]
Fit[Transpose[{voltages,Log[currents/1000]}],
    {1,v},v]
```

Out:

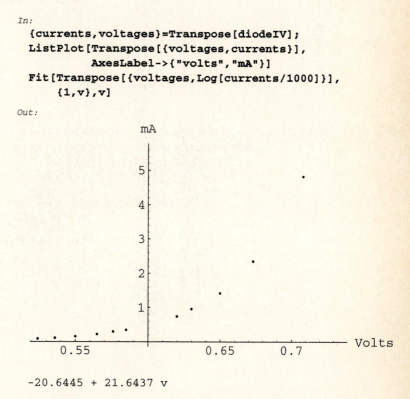

>   -20.6445 + 21.6437 v

How well does the analytic function mimic the experimental data? You can check by plotting both data and function:

In:
```
fitplot=Plot[Exp[eqn],
            {v,0.5,0.8},
            AxesLabel->{"Volts","mA"},
            PlotLabel->"Fitted data"]
```

Out:

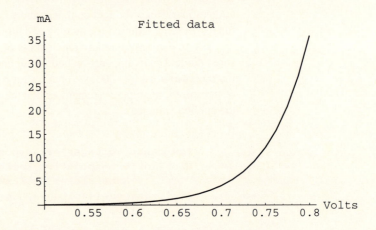

The model can be compared with the original data most easily by using the **Show** command to plot both data (held in the variable **dataplot**) and the modeled data (**fitplot**) on the same graph:

In:
```
Show[dataplot,fitplot]
```

Out:

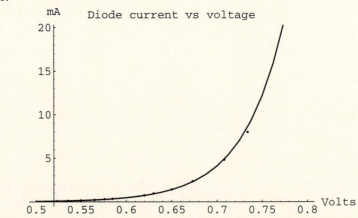

Now you have a function that gives the current through a diode as a function of the applied voltage. You can also develop current–voltage relationships using theoretical functions, of course. For general circuit behavior modeling, theoretical forms may be more suitable because they are not device specific.

As an example, take the same problem of defining a theoretical function for the current passed by a diode as a function of the voltage across the diode. This function takes a simple exponential form and uses some of the physical constants supplied by *Nodal* — the electronic charge and Boltzmann's constant; you can define your own constants using the normal procedure for user-defined functions. Note that the charge is defined (correctly) as negative but that we require its absolute value in the function. Although the diode's current is temperature dependent, we do not include temperature in the argument list; in practice, **io** is very device dependent and can range from $\approx 10^{-6}$ A to $\approx 10^{-12}$ A. You can plot **idiode** using linear or logarithmic scales for the current. (The **\n** in label strings forces the remainder of the string to flow onto a new line.)

```
In:
  io=10^-9;   (* Amps *)
  temp=300;
  Plot[ io Exp[Abs[qElectron v/
                      (kBoltzmann temp)]],
      {v,0,0.8},
      PlotRange->{0,1},
      AxesLabel->{"Volts","Amps"},
      PlotLabel->"Diode characteristics\n
        from theory"]

Out:
```

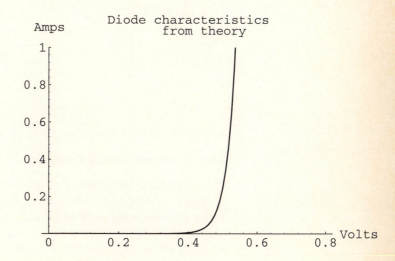

Graphics`Graphics`, one of the standard packages supplied with *Mathematica*, contains the function **LogPlot**. **LogPlot** works in the same way as **Plot** except that the *y*-axis is displayed in logarithmic format. (Another function **LogLogplot** displays both *x*- and *y*-axes in logarithmic format.) To use **LogPlot**, you first have to load the package that contains it and then you can plot the diode's current on a log scale:

In:
```
<<Graphics`Graphics`

LogPlot[idiode[v],{v,0.2,0.4},
        AxesLabel->{"Volts","Amps"},
        PlotLabel->"  Diode characteristics
                   \n   from theory
                   \n   (Log plot)"]
```

Out:

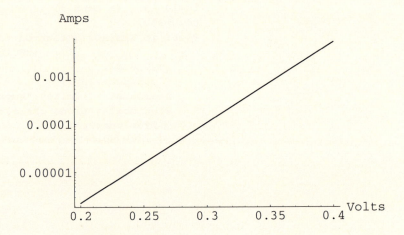

2.4.2 Nonlinear Circuit Analysis

One of the classic DC nonlinear circuit problems involves a diode: given a voltage source, resistor, and diode in series, what is the voltage across the diode? Such a problem is easily solved because you know that because the circuit elements are in series, the current through each of the elements must be the same. First, you define the parameters of the circuit and equations for the current through both the resistor (**vSupply/r - v/r**) and (in this example, using a theoretical model) the diode (**io Exp[Abs[qElectron] v/ (kBoltzmann temp)]**):

```
In:
  vSupply=10;
  r=10 kOhm;
  io=10^-9;
  temp=300;
```

You then use **FindRoot** to solve the resulting equation.

Toolbox

Finding Zero-Points or Equalities and Minima in Equations

FindRoot [*leftSide==rightSide*, {*var, startValue*}] will find the value of *var* (starting at *var=startValue*) at which the equation's *leftSide* and *rightSide* are equal. Setting either *leftSide* or *rightSide* equal to zero or omitting it will result in finding a local root in the remaining equation.

FindMinimum[*function*, {*var, startValue*}] will find the value of *var* (starting at *var=startValue*) at which the *function* is at a local minimum near *startValue*.

Because the currents through the diode and the resistor are the same, their difference must be zero and you can find the voltage at which a zero value occurs by using **FindRoot**. You must give a starting value for the iterative procedure, and here we have used **v**=0.5, a reasonable guess:

```
In:
  FindRoot[(vSupply/r - v/r)-
                  (io Exp[Abs[qElectron] v/
                              (kBoltzmann temp)]),
            {v,.5}]
Out:
  {v -> 0.356537}
```

Thus the circuit will rest in equilibrium with 357 mV across the diode and 9.643 V across the 10 kΩ resistor. Note that **FindRoot** returns a rule; the value

0.356537 is not assigned to **v**. Unless you have already assigned a value to a variable named **v**, *Mathematica* treats **v** as a new variable if you perform any computations with **v**. If you want to use **v** with the value calculated by **FindRoot**, you must use the replacement operator, **/.**:

```
In:
  v+1
```

```
Out:
  1 + v
```

```
In:
  v+1 /. %%
```

```
Out:
  1.35654
```

As another example of nonlinear circuit solving, we bias a field-effect transistor (FET) after empirically determining device I_{DSS} and V_P.

The FET is widely used in many areas of electronics — in op-amps, as a discrete device in audio- and radio-frequency amplifiers, and, particularly, in digital circuitry where it is the active element in all CMOS integrated circuits. For the FET to function as an amplifier, it has to be biased correctly. To bias the FET, you must know its DC characteristics and ensure that, in a circuit, the biasing is correct — you can perform the calculations for both these requirements using *Mathematica*.

Because all FET types have a certain commonality in electrical behavior, we will look at only one type, an N-channel JFET, to see how the DC circuit parameters can be calculated using *Mathematica*. We show the FET voltages and currents in Figure 2.8 and, in Figure 2.9, a schematic of the relationship between the current through and the voltage across the device for different voltages applied between the gate and the source. The voltage control of channel current can be clearly seen — as the gate-source voltage becomes more negative, the current (regardless of the magnitude of the drain-source voltage) decreases. At a sufficiently negative gate-source voltage, the current is effectively cut off. The voltage at which I_{DS} ceases is known as the "pinch-off" voltage, V_p, and it is an important parameter in the FET's characteristics. V_p is very device dependent: V_p varies a lot between device types as well as between devices of the same type.

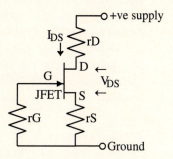

Figure 2.8   FET voltages and currents.

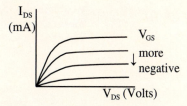

Figure 2.9   FET current-voltage characteristics.

In Figure 2.9, the FET characteristic curves show two behavioral regions: the saturated and the resistive regions. In the saturated region, the FET's channel current is determined only by the gate-source voltage: The current is independent of the drain-source voltage, within operational limits. Note that the current passed remains independent of V_{DS} as the gate-source voltage approaches the pinch-off limit. When the gate is tied directly to the source (the gate-source voltage then being zero), the current passed through the channel is known as I_{DSS} and is normally specified at a known drain-source voltage. When the FET is operated within the saturation region [Northrop90], you can calculate the channel current at a given gate-source voltage from the convenient, but approximate, model

$$I_{DS} = I_{DSS} (1 - V_{GS}/V_P)^2$$

We also show the other region of the FET's characteristic plot, the resistive region, in Figure 2.9. In the resistive region, the current in the channel is dependent on the drain-source voltage: The characteristic curves measured at different gate-source voltages are nearly straight, mimicking different values of resistors. The value of the drain-source resistance is dependent on the gate-source voltage. The boundary region that separates the saturated and resistive regions is not used.

In summary, the FET can be used either as a voltage-controlled current source or as a voltage-controlled resistor, the type of behavior obtained being dependent on the magnitude of the drain-source voltage; I_{DS} and V_{DS} both depend on the two constants I_{DSS} and V_P .

To find I_{DSS} and V_P (=V_{GS}(off)) for the junction FET, you can use the same basic procedure that you used to determine an analytical expression for diode characteristics. You obtain a list of {V_{GS}, I_{DS}} pairs, **fetpt**, then fit an analytic function:

In:
```
fetpt={{-2.2, 0.06}, {-2.17, 0.07},
       {0, 8.6},{-1.78, 0.73},
       {-1.65, 1.06}, {-1., 3.46},
       {-0.19, 7.6}};

fetdataplot=
    ListPlot[fetpt,
    PlotRange->All,
    AxesLabel->{"Vgs","Ids(mA)"},
    PlotLabel->"FET characteristics"]
```

Out:

FET characteristics

In:
```
fetfit=Fit[fetpt,{1,v,v^2},v]
```

Out:

$$8.69304 + 6.45237 \ v + 1.13715 \ v^2$$

You can determine, by inspection, that I_{DSS} (when the gate-source voltage is zero) equals 8.7 mA and, using **FindMinimum**, that the pinch-off voltage, V_P, equals –2.8 V and the drain current at pinch-off is –0.4 mA:

In:
```
FindMinimum[fetfit,{v,-3}]
```

Out:
```
{-0.459918, {v -> -2.83708}}
```

When you use any computer-generated fit, it is always a good idea to look at what has been produced and to compare carefully the fit with the original data. The drain current cannot be negative in reality — modeling the drain current by a quadratic function is only an approximation and the process of determining an analytic fit has been misled by the inaccuracies in our empirically determined data. The function we have determined is quite a good fit to the data in the range $-2.2 < V_{GS} < 0$.

In:
```
fetfitplot=Plot[fetfit,{v,-3,0},
                AxesLabel->{"Vgs","Ids"},
                PlotLabel->"FET IV function"]
```

Out:

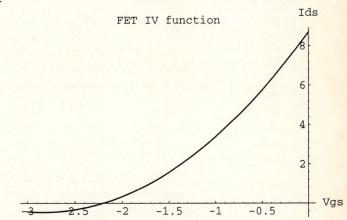

In:
```
Show[fetfitplot,fetdataplot,
     PlotRange->{{-3,0},{0,10}},
     AxesLabel->{"Vgs","Ids (mA)"},
     PlotLabel->"FET data and model fit"]
```

Out:

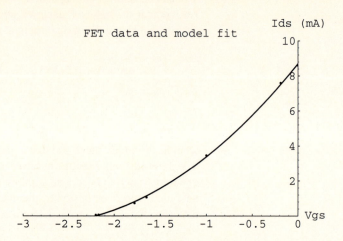

FET data and model fit

In our determination of V_p, it might have been better to find the value of V_{GS} that made I_{DS} zero, rather than to find the minimum of the $I_{DS}(V_{GS})$ function:

In:
```
FindRoot[fetfit,{v,0.5}]
```

Out:
```
{v -> -2.20112}
```

Once you know V_P and I_{DSS}, and given the drain current required, you can compute the bias values for the FET. First, specify the pinch-off voltage (**vp**), the desired drain current (**iRequested**), and **iDSS**.

In:
```
vp=-2.5;
iDSS=0.008;
iRequested=0.001;
```

Then, you need to define an expression for the drain current as a function of **vgs**. If you subtract the desired drain current, you can use **FindRoot** to numerically compute the required value of **vgs**.

In:
```
FindRoot[iDSS (1-vgs/vp)^2 - iRequested,
         {vgs,-1}]
```

Out:
```
{vgs -> -1.61608}
```

You can self-bias the FET by placing a 1.6 kΩ resistor between ground and the FET's source; **vgs** will then equal −1.6 V for your chosen drain current of 1 mA.

The gate is effectively at ground because very little current flows through the gate resistor and so, by Ohm's law, the voltage across that resistor is nearly zero.

2.5 Summary

In this chapter, you have seen how

- to calculate voltages and currents in DC circuits by applying rules and solving equations (either directly from a circuit netlist using *Nodal* or from the manual application of Kirchhoff's laws using matrices)

- to write user-defined functions

- to provide time-saving utility functions

- to import experimental data into *Mathematica*

- to find functional forms for component behavior and subsequently use them to determine circuit behavior in nonlinear circuits.

2.6 Exercises

2.1 A circuit contains 12 equal-valued resistors connected so that they make a wire-frame model of a cube (that is, one resistor along each side of the cube, each vertex connected to three resistors). Determine the resistance between a corner and (a) one of its closest neighbors and (b) its farthest neighbor.

2.2 A Schmitt trigger powered from a 1 V supply is required to monitor an analog voltage and to switch off when the voltage drops below 5.600 V. It must switch on again when the voltage rises above 5.610 V. Design a circuit to meet this specification, assuming you have a supply of perfect voltage comparators. What happens when the supply voltage drops by 1 V?

2.3 Write a user-defined function to help you calculate the power dissipated by any of the resistors in Exercises 2.1 and 2.2.

2.4 A T-pad attenuator is to be fed by a 75 Ω source and is to drive a 40 Ω load. Design an asymmetric pad that has 10 dB attenuation and whose input impedance matches the source impedance and whose output impedance matches the load impedance.

2.5 Write a user-defined function to calculate the current through a silicon diode as a function of both temperature and applied voltage. Plot the current for two different temperatures as the voltage varies from 0 to 0.7 V. (Refer to the Appendix for help on graphics functions.)

2.6 Write a list of 10 calculations that you perform often and that would be suitable for inclusion in your own library of functions.

2.7 Choose a FET from your component store, and measure its V_p and I_{DSS}. Calculate the V_{GS} required to make I_{DS}=100 μA. Look up the FET's specification in the manufacturer's databook, and calculate the minimum and maximum values of I_{DS} (at your calculated value of V_{GS}) that would result from the minimum and maximum values quoted for both V_p and I_{DSS}.

2.8 Write a *Mathematica* program that will compute the resistor values (from the E-12 series) required to make a potential divider, given the supply and tap voltages.

2.9 Write a *Mathematica* program to compute bias resistor values for a bipolar transistor (which can be composed of germanium or silicon), including consideration of the effects of temperature on device operation.

2.7 References

[Adby80] Adby, P.R., *Applied Circuit Theory*, Ellis Horwood, Chichester, United Kingdom, 1980.

[Delaney69] Delaney, C.F.G., *Electronics for the Physicist*, Penguin Books, Harmondsworth, United Kingdom, 1969.

[Northrop90] Northrop, R.B., *Analog Electronic Circuits: Analysis and Applications*, Addison-Wesley, Reading, Massachusetts, 1990.

[Vlach83] Vlach, J., Singhal, K., *Computer Methods for Circuit Analysis and Design*, Van Nostrand Reinhold, New York, 1983.

[Wolfram91] Wolfram, S., *Mathematica — A System for Doing Mathematics by Computer*, Addison-Wesley, Redwood City, California, 1991.

CHAPTER 3

Small-Signal Circuits I: Introduction

This chapter begins our study of small-signal AC circuit analysis with *Mathematica*. In the circuit analysis context, a small signal is defined as being of such a magnitude that neither the linearity of the circuit nor the values of components within the circuit are affected by its presence. Determining the magnitude and phase of AC signals is a fundamental task in AC circuit analysis. In this chapter, we show how to use *Mathematica* and *Nodal* to calculate magnitude and phase of AC signals, to obtain frequency-domain transfer functions, to use models of circuits containing active devices, and to compute the dynamic properties of components.

3.1 Magnitude and Phase Calculation in RLC Circuits

Voltage and current are proportional to each other in purely resistive linear circuits. In circuits that contain capacitors and inductors, the complex impedances of those components can cause a phase difference (a lead or lag) between voltage and current. A vector — representing both the magnitude and the phase difference of a signal — can be expressed in a number of ways: as a complex number or as a relative magnitude and angle. The angle can be in radians or in degrees; the relative magnitude of voltages (or currents) can be expressed as an arithmetic ratio or in decibels. *Mathematica* can help you to calculate voltage vectors at circuit nodes and also to display the result in whatever format you want.

In Section 3.1.1 we show how you can analyze very simple circuits by applying Ohm's law in its complex form or by using *Nodal*'s **NodalAnalyze** function. In Sections 3.1.2 and 3.1.3, we discuss how you can use *Mathematica* to solve the equations that result from the application of Kirchhoff's laws in both mesh and nodal analysis.

3.1.1 Elementary Technique

The simplest technique for solving AC circuit problems is to apply Ohm's law in its complex form. For example, in a simple resistor–capacitor circuit — a 100 kΩ resistor and 100 pF capacitor in series — that is subjected to an AC sinusoidal voltage, you can compute the ratio of the output voltage (at the junction of the resistor and capacitor) to input voltage at a known frequency (10 kHz, in this example) using the complex form of Ohm's law. You can use *Mathematica* alone for this task or *Nodal*'s **NodalAnalyze** command:

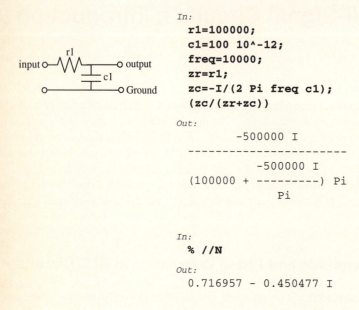

```
In:
 r1=100000;
 c1=100 10^-12;
 freq=10000;
 zr=r1;
 zc=-I/(2 Pi freq c1);
 (zc/(zr+zc))

Out:
                 -500000 I
         -----------------------
                        -500000 I
         (100000 + ---------) Pi
                        Pi


In:
 % //N

Out:
 0.716957 - 0.450477 I
```

If you perform the equivalent calculation using *Nodal*, you can go directly from a netlist description of the circuit to the result:

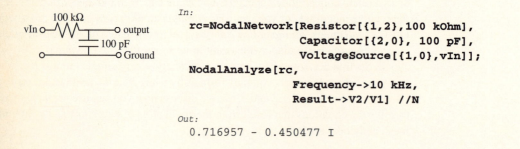

```
In:
 rc=NodalNetwork[Resistor[{1,2},100 kOhm],
                 Capacitor[{2,0}, 100 pF],
                 VoltageSource[{1,0},vIn]];
 NodalAnalyze[rc,
                 Frequency->10 kHz,
                 Result->V2/V1] //N

Out:
 0.716957 - 0.450477 I
```

Both *Mathematica* and *Nodal* give the voltage ratio as a vector in the complex plane.

You can use the **Abs** and **Arg** functions of *Mathematica* to compute the length of the voltage vector and the phase change in radians.

Toolbox

Elementary Complex Number Functions in *Mathematica*

Use **I** as $\sqrt{-1}$: A complex number z is represented in the form $x + y$ **I**. For a complex number, z, you can extract its real and imaginary parts using **Re**[z] and **Im**[z].

The absolute value (|z|), argument (ϕ, where $z = |z|e^{i\phi}$), and conjugate (z^*) can be found using the functions **Abs**[z], **Arg**[z], and **Conjugate**[z], respectively.

```
In:
   Abs[.716957-0.450477 I]
Out:
  0.846733
```

```
In:
   Arg[.716957-0.450477 I]
Out:
  -0.560982
```

If you wish to reformat the complex number form of the output-to-input voltage ratio of the network, you can write your own user-defined function or use one of *Nodal*'s utility functions.

Here is an example of a user-defined function, called **complexPolar**. This function takes as its argument a single complex number, **x**, and returns a two-element list containing the magnitude and phase (in degrees) of **x**. We have used the postfix form of **N** to force **complexPolar** to return the numeric form of its answer, whenever possible:

```
In:
   complexPolar[x_Complex]:={Abs[x],
                              Arg[x] 180/Pi}//N
```

```
In:
   complexPolar[1+I]
Out:
  {1.41421, 45.}
```

Other utilities in *Nodal* will give you a description of the vector in terms of polar coordinates, $re^{I\Theta}$ or (r,θ), rather than the Cartesian form $(x, y$I); the **Angle** utility in *Nodal* gives the phase in degrees:

```
In:
    ComplexToPolar[1 + 2 I]
```

Out:

$$\text{Sqrt[5] E}^{\text{I ArcTan[1, 2]}}$$

```
In:
    ComplexToPolarDegree[.716957 - 0.450477 I]
```

Out:

$$\{0.846733, \frac{-0.560982}{\text{Degree}}\}$$

```
In:
    Angle[.716957-0.450477 I]
```

Out:

    -32.1419

Mathematica handles equations symbolically whenever it can, and it left the answer from **ComplexToPolarDegree** in a symbolic form because an intermediate value has been referred to by name (that is, **Degree**). You can determine any answer's numerical value (if it has one) by using the enumerate function, **N**:

```
In:
    ComplexToPolarDegree[.716957 - 0.450477 I]//N
```

Out:

    {0.846733, -32.1419}

```
In:
    ComplexToPolarDBDegree[.716957 - I 0.450477]//N
```

Out:

    {-1.44507, -32.1419}

You can use the inverse-dB utility **ArcDB** to convert from dB to an arithmetic ratio:

```
In:
    ArcDB[-1.44507]
```

Out:

    0.846733

Should you require dB for a power ratio, rather than a voltage or current ratio, then *Nodal* provides the **DBP** utility. **DBP**[x] returns $10 \, \text{Log}_{10}(x)$, or **10 Log[10,**x**]** in *Mathematica* notation.

One practical application of AC analysis techniques is the determination of the effects of stray capacitance on circuit performance. A perfect resistive divider should have frequency-independent behavior. In practice, frequency independence is not achieved because each of the resistors in the divider will have a small-valued stray capacitance. What is the effect of this stray capacitance at high frequencies?

First, what is meant by high frequency? You can define high frequency as frequencies at which the assumption $Z(c) \ll Z(r)$ is no longer valid — that is, the impedance of the resistor's stray capacitance degrades the divider's performance.

You can define a *Mathematica* function, **z**, to calculate the impedance of a capacitor, where **I** is $\sqrt{-1}$ (equivalent to j in many engineering texts), **Pi**=π, **freq** is the frequency used, and **cap** is the capacitance (in farads). You can then use **FindRoot** to calculate the frequency at which the magnitude of the stray capacitor's impedance equals the resistance of the resistor. If *Mathematica* can find a symbolic derivative of the function upon which **FindRoot** is to operate, you need only give a single first-guess value. However, in this example, because we want the magnitude of the capacitor's impedance, we have used the **Abs** function for which no symbolic derivative is defined, and so we must give **FindRoot** a list of two starting values:

```
In:
   z[cap_, freq_]:= -I/(2 Pi freq cap) //N
   FindRoot[Abs[z[1 pF,frequency]]==10^5,
            {frequency, {1 kHz,1 GHz}}]

Out:
                                         6
   {frequency -> 1.59155 10 }
```

You can calculate the effect of a stray capacitance of 1 pF associated with each resistor by defining a new circuit and then analyzing it using Kirchhoff's laws or **NodalAnalyze**. Because the computation is frequency dependent, **NodalAnalyze** includes the rule option, **Frequency->f**, so the variable **f** is used as the variable for the frequency:

```
In:
   rDivider2 = NodalNetwork[
                    VoltageSource[{1,0},vIn],
                    Resistor[{1,2},r1],
                    Resistor[{2,0},r2],
                    Capacitor[{1,2},1 pF],
                    Capacitor[{2,0},1 pF]];
   rD2Out = NodalAnalyze[rDivider2,
                    Result->V2/V1,
                    Frequency->f] //Simplify
```

Out:

```
         -I                    1
    ------------ f Pi -   --
    500000000000            r1
    ------------------------------
         -I               1       1
    ------------ f Pi -   --  -   --
    250000000000            r1      r2
```

Inspection of this equation shows you that for **f**=0, the output of the circuit will be **r2/(r1+r2)** and, when **f** is very large, the value will approach 1/2. When Pi f r1 r2 = 500000000000 r2, this resistive divider will be failing. For the component values used in this example, a little mental arithmetic reveals that failure will occur around **f**=(5/π) MHz, a frequency that agrees with the answer from **FindRoot**. However, it is much easier to obtain an understanding of the circuit's behavior by studying a graph. You can show graphically the effect of the stray capacitances by plotting **rD2Out** over a range of frequencies, say from 10 Hz to 10 GHz. The easiest way to generate a plot of **rD2Out** is to use the infix form of **ReplaceAll** (**/.**) to replace the symbolic variables **r1**, **r2**, and **f** with their numeric values in **rD2Out**. Note that we have placed all the replacement rules in a list and used *Nodal*'s **DB** function to return the value of **rD2Out** in decibels. By setting **f->10^freq** and specifying the plotting variable **freq** to range from 1 to 10, we cause the value of **rD2Out** to be plotted from 10 Hz to 10 GHz:

In:
```
Plot[DB[Normal[rD2Out][[1,2]] /. {r1->100 kOhm,
                                  r2->1 kOhm,
                                  f->10^freq}],
     {freq,1,10}, PlotRange->{-50,0},
     AxesLabel->{"Log(f)","dB"}]
```

Out:

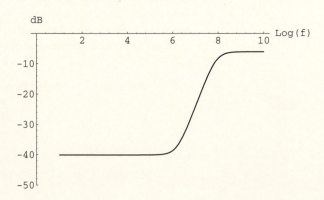

The graph shows how the attenuation of the potential divider becomes increasingly governed by the stray capacitance of the resistors as the frequency is increased. In this example, the divider is operating within specification only at frequencies up to ~100 kHz; when a frequency of 100 MHz is used, the divider ratio is completely dominated by the action of the capacitors.

Mathematica can help you design a better potential divider. Given that it is not possible to have no stray capacitances, is it possible to choose values for larger-valued capacitors in parallel with the resistors that would result in less performance degradation? First, we generalize the problem by replacing the stray capacitances with impedances — **z1** and **z2** — in parallel with the resistors **r1** and **r2**. Second, because we want to keep the problem totally symbolic, we need to use **Clear** to deassign any numeric value assigned, during our numeric example, to components. Third, we reanalyze the circuit and then use **Solve** to find values for the impedances that maintain the division ratio of our new divider equal to **r2/(r1+r2)**.

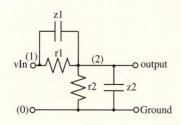

```
In:
  Clear[r1,r2,c1,c2];
  rDivider3=NodalNetwork[VoltageSource[{1,0},vIn],
                         Resistor[{1,2},r1],
                         Resistor[{2,0},r2],
                         Impedance[{1,2},z1[f]],
                         Impedance[{2,0},z2[f]]];
  rD3Out=NodalAnalyze[rDivider3,
                      Result->V2/V1,
                      Frequency->f];
Normal[rD3Out][[1,2]]
```

```
Out:
```

$$\dfrac{-z2[f] - \dfrac{z1[f]\ z2[f]}{r1}}{-z1[f] - z2[f] - \dfrac{z1[f]\ z2[f]}{r1} - \dfrac{z1[f]\ z2[f]}{r2}}$$

```
In:
  Solve[%==(r2/(r1+r2)),{z1[f],z2[f]}]
```

```
Out:
```

$$\{\{z1[f] \rightarrow \dfrac{r1\ z2[f]}{r2}\}\}$$

The performance of the new divider will be maintained if the impedance ratio `z1[f]/z2[f]` equals the ratio of the resistors, `r1/r2`. Implementing the impedance components using capacitors, you would therefore choose capacitors such that `c1/c2 = r2/r1`. You can check that the performance of the new divider is indeed frequency independent by seeing whether the result from **NodalAnalyze** contains the frequency variable, **f**:

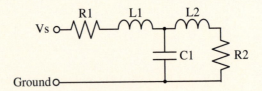

```
In:
    rDivider4=NodalNetwork[VoltageSource[{1,0},vIn],
                    Resistor[{1,2},100 kOhm],
                    Resistor[{2,0},1 kOhm],
                    Capacitor[{1,2},1/100 nF],
                    Capacitor[{2,0},1 nF]];
    NodalAnalyze[rDivider4,
                Result->V2/V1,
                Frequency->f]//Simplify

Out:
    1
   ---
   101
```

3.1.2 Mesh Analysis

You can tackle more complicated passive RLC circuits using *Mathematica* and *Nodal*. Using *Nodal*, you can proceed directly from a netlist to results but, if you want to perform your analysis using only *Mathematica*, you need to apply Kirchhoff's voltage and current laws. Either of these two laws allows you to completely analyze a circuit using a set of equations that describes each loop or node in the circuit. Figure 3.1 shows a filter between a voltage source with internal resistance, **R1**, and a load, **R2**. How do you find the voltage across **R2**, **VR2**, in terms of the source voltage, **Vs**, and the transfer function of the circuit, **VR2/Vs**?

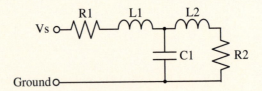

Figure 3.1   Five-component filter network.

Every set of components that is connected in a continuous loop forms a mesh. Currents can be assigned to two of the meshes in the circuit of Figure 3.1

to give **i1** and **i2**. Because Kirchhoff's voltage law states that the sum of the voltage drops around a loop must equal zero, if you collect equations that describe the voltage around a loop and solve them simultaneously, you will arrive at a solution providing that the number of unique loops equals the number of unknowns. In *Mathematica* notation,

```
Sum[VoltageDropsAroundALoop[i], {i,1,nLoops}] == 0
```

Note that voltage drops indicate a reduction in the voltage in the direction of the current. **mesh1** begins with a voltage rise due to **Vs** and then has voltage drops due to **R1**, **L1**, and **C1**:

```
mesh1 = -Vs + VR1 + VL1 + VC1;
mesh2 = -VC1 + VL2 + VR2;
```

To solve these equations, you write them in terms of the two mesh currents, so the two mesh currents become the two unknowns in the two equations. You can do this by setting up some substitution rules and making the assignment **p** = **2π f I**, and then solving the equations subject to those rules:

```
In:
meshRules = {VR1 -> i1 R1,
             VL1 -> i1 p L1,
             VC1 -> (i1-i2)/(p C1),
             VL2 -> i2 p L2,
             VR2 -> i2 R2};

soln = Solve[{mesh1,mesh2} == {0,0} /. meshRules,
             {i1,i2}]
```

```
Out:
                                                 2
{{i1 -> (C1 Vs p (1 + C1 R2 p + C1 L2 p )) /

                        2          2    2           2
(C1 R1 p + C1 R2 p + C1 L1 p + C1 L2 p + C1  R1 R2 p +

      2        3    2      3    2      4
    C1  L2 R1 p + C1  L1 R2 p + C1  L1 L2 p ),

i2 -> (C1 Vs p) /
                        2          2    2           2
(C1 R1 p + C1 R2 p + C1 L1 p  + C1 L2 p  + C1  R1 R2 p +

      2        3    2       3    2      4
    C1  L2 R1 p  + C1  L1 R2 p  + C1  L1 L2 p )}}
```

Note that a fourth-order denominator in **p** is present because a common **p** term exists in both the numerator and denominator of **i1** and **i2**. **VR2/Vs** is the desired transfer function and because **VR2** = **i2 R2** and **i2** is in terms of **Vs**, you can solve for the transfer function:

```
In:
  transferFunction =
  Simplify[i2 R2/Vs  /.
                soln][[1]]//ExpandAll//Cancel
```

```
Out:
                                                                2
    R2 / (R1 + R2 + L1 p + L2 p + C1 R1 R2 p + C1 L2 R1 p  +

                    2            3
          C1 L1 R2 p  + C1 L1 L2 p )
```

Note that the **[[1]]** index removed the braces from the result as formatted in **soln**. We have also used **ExpandAll** and **Cancel** to tidy the form of the answer.

You now have a cubic equation in the denominator because of the three reactive elements in the circuit. To get a feel for the performance of this network, you can plot its frequency-domain response. First, with some loss in generality, you can parameterize the network in terms of the Q of its components. Because the Q of the components is proportional to the energy stored, the parameterization will allow you to see how the network responds as more energy is stored in the components. Given the following rules, you can set up a parameterized transfer function for plotting. (We make **R1** equal to **R2**, for convenience.)

```
In:
  parameterRules = {L1->q R2/(2 Pi),
                    C1->1/(q R2 2 Pi),
                    L2->q R2/(2 Pi),
                    R1->R2,
                    p->2 Pi I f};
  pTF = Simplify[transferFunction /. parameterRules]
```

```
Out:
                              q
  -----------------------------------------
                  2            2        3 2
  I f + 2 q - 2 f  q + 2 I f q  + -I f  q
```

Note that **R2** is no longer a factor because it was absorbed into the **q** term. We can now plot **pTF** for three different values of **q**; we have to use **Evaluate** to force *Mathematica* to make a list of three functions with the numeric values 0.1, 1, and 10 substituted for **q**. If **Evaluate** were not used, **Plot** would hold its first argument in symbolic form and so compute an unplottable list of three numbers for each value of **f**. Because the plot label is rather long, we have forced it onto two lines by inserting a new line character (**\n**) into the text string:

```
In:
    Plot[Evaluate[DB[pTF /. q->{0.1,1,10}]],
        {f, 0, 5},
        AxesLabel->{"f","dB"},
            PlotLabel->"transfer\nfunction"]
Out:
```

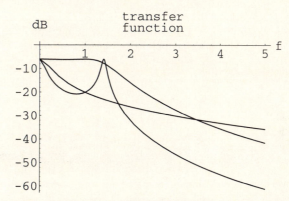

From the graphs, you can see that the low-frequency loss is 6 dB and the high-frequency loss is a function of **q**. The **q**=0.1 curve has a rapid monotonic roll-off at DC but a slower roll-off at higher frequencies. The **q**=1.0 curve is fairly flat out to **f**=1.3 and then rolls off rapidly. The **q**=10 curve has a sharp resonance near **f**=1.4 and the fastest high-frequency roll-off. Because of the **q** relationships, the capacitance is increased at low **q** and the inductance is decreased, making the circuit dominated by capacitance at low **q**. As **q** is increased, the inductance first balances the capacitance and finally overcompensates it to produce a resonance. Note that for the three values of **q**, *Mathematica* has merely had to reevaluate, not rederive, the transfer function **pTF**.

3.1.3 Nodal Analysis

The dual of mesh analysis is nodal analysis. Nodal analysis is the easiest to construct with a computer, so it forms the basis of most computer-aided analysis programs. Some circuits cannot be described by nodal analysis, so you have to use a combination of nodal and mesh analysis [Vlach83].

Figure 3.1 shows the third-order circuit from a nodal viewpoint. Here, the system of equations is based on the sum of currents into each node being zero. In *Mathematica* notation,

```
Sum[CurrentsIntoANode[i], {i,1,nNodes}] == 0
```

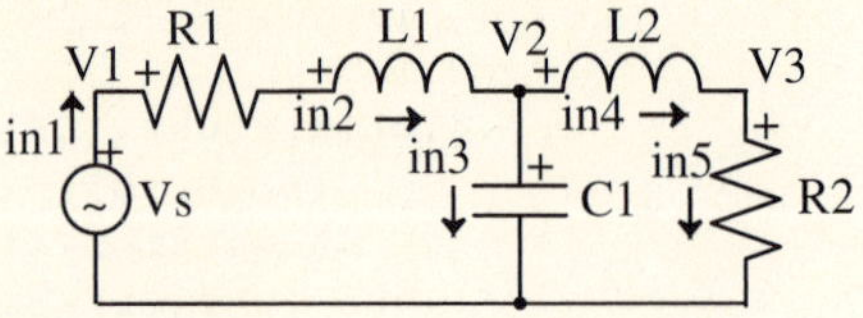

Figure 3.2   Circuit with three nodes and labeled currents.

Figure 3.2 shows a circuit with three nodes and a ground node. The ground node is not labeled; we assume it to be at zero potential, so each node voltage is absolute rather than relative. Taking currents into a node as positive and out of a node as negative, you can formulate the node currents as follows:

```
node1 = in1 - in2;
node2 = in2 - in3 - in4;
node3 = in4 - in5;
```

To find the solution, you can write these equations in terms of the three node voltages by setting up some substitution rules in a manner analogous to the mesh analysis technique:

```
In:
nodeRules = {in1 -> (Vs-V1)/R1, in2 -> (V1-V2)/(p L1),
             in3 -> V2 p C1, in4 -> (V2-V3)/(p L2),
             in5 -> V3/R2};
```

The node solution has the three node voltages as unknowns. You need **v3** in terms of **Vs** — obtainable directly from the rules returned by **Solve**. (Note that we use the **[[1]]** index to extract the expression from the braces.)

```
In:
v3Soln =
  V3 /. Solve[{node1,node2,node3} == {0,0,0} /.
              nodeRules, {V1,V2,V3}][[1]]
Out:
 -((L1 L2 R2 Vs p) /
                                        2      2          2  2
     (-(L1 L2 R1 p) - L1 L2 R2 p - L1  L2 p  - L1 L2  p  -

                       2                2    3         2          3
        C1 L1 L2 R1 R2 p  - C1 L1 L2  R1 p  - C1 L1  L2 R2 p  -

                2  2  4
        C1 L1  L2  p ))
```

This answer looks more complicated than the previous solution for **i2**. The only difference should be the **R2** multiplier; the **p** term in the numerator will cancel the fourth-order term in the denominator to give the third-order transfer functions you expect from this circuit. You can divide this solution by **Vs** and simplify it to get the same transfer function as before:

In:
```
nodeTransferFunction = Simplify[v3Soln/Vs]
```
Out:

$$
R2 \; / \; (R1 + R2 + L1\,p + L2\,p + C1\,R1\,R2\,p + C1\,L2\,R1\,p^{2} +
$$
$$
C1\,L1\,R2\,p^{2} + C1\,L1\,L2\,p^{3})
$$

The ability to check solutions in both symbolic and numerical forms is extremely useful, and it is important to check all results generated by a computer. You can easily check whether this is the same solution as before by testing for equality:

In:
```
nodeTransferFunction == transferFunction
```
Out:
```
True
```

The above circuit was simple to analyze. As the circuit gets larger, however, the number of nodes or meshes increases and the analysis becomes more tedious. Thus the user needs easier ways to specify and present the problem in a form the computer can solve. The easiest way for you to specify the problem is to draw a diagram and specify the node you are interested in. However, text-based analysis tools, such as *Mathematica*, cannot currently use diagrams as input, so we have to use the second easiest way to describe circuits — a netlist of components and connection points.

A netlist description has several advantages over a diagram: It can be loaded directly into a system of equations to be solved, it is faster to describe, and it can serve as a common language between CAD tools. The disadvantage of a netlist is that if you or anyone else wants to understand it, you must first convert it into a circuit schematic.

A simple nodal analysis program that first translates a circuit netlist into a matrix description is described in Chapter 4.

3.2 Small-Signal Analysis Techniques

So far, our look at AC signals in circuits has involved only passive components. Because much of the rest of the book is about small-signal analysis when active components are present, we feel that it is important to state what assumptions and modifications you must make during analysis.

Small-signal analysis first assumes that circuit elements are not perturbed by the signal passing through them and that the operating conditions (for example, semiconductor biasing) of the circuit therefore are not altered. By comparison, nonlinear analysis recomputes circuit operating conditions for every timestep during the analysis. Another aspect of small-signal analysis is that dynamic characteristics of components are required. We discuss both these topics in Sections 3.2.1 and 3.2.2.

3.2.1 Preparation for Analysis

In Chapter 1, we described a simple scenario where an amplifier with a gain of 100 is fed with 10 μV and 1 V AC signals. By small-signal analysis, the output signal is, of course, 1 mV and 100 V, respectively: Small-signal techniques assume that the amplifier is at all times operated in its linear region. If you know that the amplifier is powered from a 9 V battery, then it should be obvious that a 1 V input signal will not give 100 V out!

If you were to carry out a full nonlinear AC/DC analysis, the output from your CAD program would show that the amplifier's output started to clip long before 100 V was reached. But such full analysis is nugatory if you know the circuit you are analyzing will always be operated in its linear region. Because nonlinear AC /DC analysis is very compute-intensive, you will find it quicker to resort to small-signal analysis where possible.

When you decide to design a circuit, your first task might be to ask several questions and to think about what CAD tools you might want to use.

What circuit configuration shall I use?

By reducing a circuit to its simplest and most idealized form, you can investigate the fundamental characteristics of a given configuration. Questions such as How sensitive is the input impedance to frequency? are best answered at an early stage of design. Small-signal analysis is of great value in answering such questions.

How are the active components powered and biased?

Mathematica can help you solve the DC equations that determine the values of bias components, power supply loading, and component power dissipation.

Does the circuit operate in a linear or in a nonlinear regime?

By using *Mathematica* to solve the nonlinear equations, you can find out whether the linearity assumption still holds.

How will, say, device capacitances affect performance?

Component models — from *Nodal*'s library or made up to your own requirements — can incorporate small inductances and capacitances. Their effect on circuit performance can be found numerically or symbolically using *Mathematica*.

Small-signal analysis can help you with two important areas of design: choice of the fundamental circuit configuration and the second-order problems of nonideal device characteristics. The DC and linear/nonlinear design problems are tractable with CAD — using either a general-purpose tool like *Mathematica* or a dedicated numerical system like SPICE.

When you undertake small-signal analysis, you are not concerned with the DC state of the circuit. Instead, you look at the circuit purely from the point of view of an AC signal, an approach that requires you to think about the circuit in a different way.

You can still think of the circuit as a network of components joined together at nodes; the effective value of each component is its AC impedance. For example, Figure 3.3 shows three versions of the same common-emitter transistor amplifier.

Note that the power supply has been replaced by a short circuit. Perfect voltage sources offer zero impedance to AC signals; perfect current sources have an infinite AC impedance. In practice, voltage and current sources will have some nonzero and finite impedance, respectively, and you may find that you want to refine analysis to that level of detail later on.

The transistor has been decomposed into a current source ($h_{fe}i_b$) and resistor (h_{oe}) on the output side and a resistor (h_{ie}) and voltage source ($h_{re}v_{ce}$) on the input side. All the *h*-parameters are functions of the DC conditions under which the transistor is operating. For example, h_{oe} is a function of I_c and V_{ce}. When you apply small-signal analysis to the circuit, you will need to use the values of these parameters that are correct for the DC conditions under which you will use the circuit. Using the right values is important since some parameters vary strongly with the DC conditions: h_{re} changes by a factor of ten when I_c varies from 100 µA to 1 mA, for instance. Using the right values need not involve a lot of extra work: Many manufacturers include charts of device parameters (as functions of DC conditions) in their data sheets. You may also be able to use *Mathematica*'s fitting functions to create simple functions that return a parameter's value given the DC conditions.

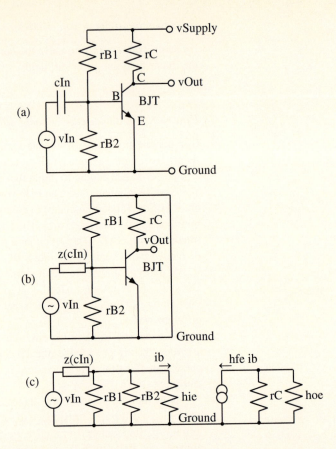

Figure 3.3 Different perspectives on the same circuit: (a) shows the normal circuit diagram, (b) shows components replaced by their AC impedance, and (c) shows the same circuit broken down into basic, discrete-function units.

3.2.2 Dynamic and Static Device Properties

When you perform small-signal analysis, often you will require the dynamic resistance of a component, not the static resistance. These two resistances are conceptually quite different, although for some devices they are always identical.

The static resistance, R, of a device (in a particular operating state) is just the voltage across the device divided by the current being passed: that is, Ohm's law. For small-signal analysis, you want to know how the current changes through the device as the result of a small change in the applied voltage — this is measured by the dynamic resistance, R_d. The difference between R and R_d is more easily understood by considering a graph showing the current passed by a device as a function of the applied voltage. Figure 3.4 shows how the current through a tunnel diode varies as the voltage across it is increased: At first the current increases, then it reaches a peak, decreases into a valley, and finally increases once again. Choosing any point on the line defines a state. For that state, $R = V/I$ and $R_d = \partial V/\partial I$, where $\partial V/\partial I$ is the reciprocal of the slope of the line.

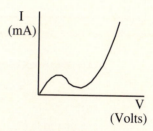

Figure 3.4   Tunnel diode characteristics.

For devices with ideal Ohm's law behavior $R = R_d$. A small number of semiconductor devices (like the tunnel diode) can have negative R_d and positive R at the same time; a negative R_d merely implies that as the voltage across the device increases, the current passed decreases (for example, in the region between the peak and the valley in Figure 3.4). A negative R_d does not imply that the device does not obey Ohm's law at the (instantaneous) operating point nor does it imply that current is flowing backward! (Sometimes, such devices are said to possess "negative resistance" but the correct phrase is "negative dynamic resistance.")

From a DC point of view, it is R that is important; in small-signal analysis, it is R_d that determines the effective resistance of the device to a small-signal passing through the device. Using an analytical form for the current versus voltage characteristic, *Mathematica* is able to compute R_d at the operating point.

For example, the simple DC resistance, **rDC**, of a diode can be calculated by dividing the voltage applied across the diode (0.356 V in this example) by the current passed:

```
In:
  io=10^-9;
  temp=300;
  rDC=0.356/(io Exp[-qElectron 0.356/
                        (kBoltzmann temp)])
Out:
  376.903
```

You can find the slope of the diode's current-voltage characteristic by differentiating the current function to obtain $\partial I/\partial V$.

Toolbox

Differentiation

D[_function_, _variable_**]** gives the partial derivative of _function_ with respect to _variable_.

```
In:
    D[io Exp[-qElectron v/
                      (kBoltzmann temp)],v]

Out:
                    -8   38.6473 v
          3.86473 10    E
```

You can then evaluate the derivative (and hence the slope of the current–voltage characteristic) at the operating point by setting **v** to 0.356:

```
In:
    rDynamic= 1/(% /. v->0.356)

Out:
    27.3943
```

You can model the diode as a resistor of value R_d that has a capacitance in parallel. The AC parameters of the diode model used by _Nodal_ are the dynamic resistance and capacitance and the parasitic bulk and contact resistance.

The diode's capacitance, C_d, can be found in the manufacturer's data sheet; small diodes used for signal handling usually have a C_d of a few picofarads.

An example of circuit analysis with dynamic and static resistances in diodes is the calculation of ripple passed through a diode; the DC behavior of the circuit can be separated from the AC behavior (an approach justified by the principles of Fourier analysis).

At an operating voltage of 0.356 V, the diode's static resistance is $\approx$377 Ω whereas its dynamic resistance is only $\approx$27 Ω . If the diode was in series with a 377 Ω resistor, the supply voltage would be shared equally between the diode and the resistor but any AC signal (ripple) on the supply would see a potential divider containing two resistors of value 377 Ω and 27 Ω. The ripple at the positive end of the diode is therefore 27/(377 + 27) of its magnitude on the supply. A simple, but incorrect, analysis might have implied that it should be roughly halved, given that the diode's static resistance is similar in magnitude to the resistance of the resistor.

Note that this technique and the functions in *Nodal* do not perform DC analysis: We assume the diode to be conducting and using the same dynamic resistance and capacitance as defined by its component specification in the netlist. Invariance of component values is a standard assumption with small-signal analysis; if you use the circuit with signals that are large enough to alter the static or dynamic component values, then this assumption — and any analysis based upon it — is invalid. For example, using such a large ripple voltage would cause the diode in the previous example to become reverse-biased and so become nonconducting or cut off. That would definitely be a subject for nonlinear analysis.

3.3 Analysis of Circuits Containing Active Devices

Nodal's component library contains several types of active devices. In this section, we show you examples of their use and show how you can find the values of internal components from manufacturer's data sheets.

3.3.1 Voltage-Controlled Voltage Sources

The simplest form of active device is a voltage-controlled voltage source (VCVS) whose output is a simple multiple of its input voltage. Such a device is commonly available as an operational amplifier. *Nodal* provides a simple voltage-controlled voltage source, as well as two operational amplifier components.

You can model a basic amplifier as a VCVS fed by a voltage source:

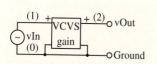

```
In:
    amp=NodalNetwork[VCVS[{1,0,0,2},gain],
                     VoltageSource[{1,0},vIn]]
    vOut=NodalAnalyze[amp,
                      Result->V2]

Out:
    gain vIn
```

All amplifiers have difficulty maintaining gain at high frequencies, so the VCVS function also allows you to specify a break-point in frequency at which a 6 dB/octave gain decrease begins. For example, you can specify a breakpoint at 1 MHz for a VCVS with a gain of 100 and plot the resulting frequency-dependent VCVS gain:

```
In:
   amp=NodalNetwork[VoltageSource[{1,0},vIn],
                    VCVS[{1,0,0,2},100,1 MHz]];
   gain=Normal[NodalAnalyze[amp,
                    Result->V2/V1,
                    Frequency->f]][[1,2]]
Plot[DB[gain /. f->10^x],
     {x,0,9},
     PlotRange->All,
     AxesLabel->{"Log(f)","Gain dB"}]
```

```
Out:
        100
     -------------
          I
   1 + ------- f
       1000000
```

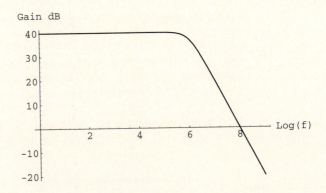

You can use *Nodal*'s Bode plot function to show how both the magnitude and the phase of the output signal vary with frequency:

```
In:
   BodePlot[NodalAnalyze[amp,
                 Result->V2/V1,
                 Step->
                    Frequency[kHz 10^Range[1,5]]]]
```

Out:

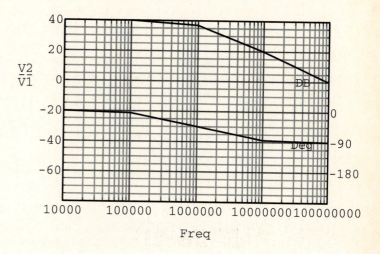

The symbolic form of an answer can be long — sometimes occupying many pages. To restrict the generality of the answer, you may want to shrink the answer by setting one or more parameters of the circuit to zero or infinity. For example, the gain of a circuit that contains a VCVS can be controlled by feedback resistors:

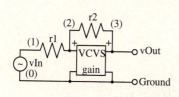

In:
```
Clear[gain];
amp2=NodalNetwork[VoltageSource[{1,0},vIn],
                  Resistor[{1,2},r1],
                  VCVS[{2,0,0,3},-gain],
                  Resistor[{3,2},r2]];
NodalAnalyze[amp2,Result->V3]
```

Out:
```
         gain vIn
    -(------------------)
        1    1    gain
     r1 (-- + -- + ----)
        r1   r2    r2
```

In:
```
Simplify[%]
```

Out:
```
        gain r2 vIn
    -(------------------)
      r1 + gain r1 + r2
```

Many amplifiers have very large (>100 dB) gains. What happens to the output voltage of **amp2** when the gain of the VCVS becomes very large? You can use *Mathematica*'s **Limit** function to evaluate an expression when one or more of the variables within the expression tend to either zero or infinity. Before you use **Limit**, you need to isolate the answer from its enclosing list structures by using **Normal**. Then, by using the rule **gain->Infinity** (**Infinity** is one of *Mathematica*'s constants), you can evaluate the circuit's performance when high-gain amplifiers are used:

In:
    Limit[Normal[%][[1,2]],gain->Infinity]

Out:

$$-\left(\frac{r2\ vIn}{r1}\right)$$

During the design phase of any circuit, if you reduce the circuit to its simplest form, you can see its behavior in its purest form, unmodified by the complicating characteristics of real electronic components. With the simple circuit **amp2**, you found that the gain is affected by two resistors — the input and feedback resistors. Once you have established the basic characteristics of your circuit, you can then use *Mathematica* to help identify the side effects of nonideal components.

For example, what is the effect of adding a capacitance between the input and the output of an amplifier? Capacitances joining the input and output of amplifiers are common and can be due to internal capacitances in transistors or to stray capacitances. They cause the well-known Miller effect (see Figure 3.5): The capacitance between the input and the output is multiplied in value by the gain of the circuit, causing a frequency-dependent fall-off of gain (**V3/V1**):

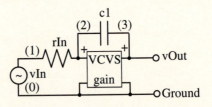

Figure 3.5   Amplifier with Miller effect.

In:
    miller=NodalNetwork[VoltageSource[{1,0},vIn],
                      Resistor[{1,2},r1],
                      VCVS[{2,0,0,3},-gain],
                      Capacitor[{2,3},c1]];

```
NodalAnalyze[miller,
              Result->V3/V1];
ampGain=Simplify[Normal[%][[1,2]]]
```

Out:

```
                     gain
-------------------------------------------
-1 - 2 I c1 f Pi r1 - 2 I c1 f gain Pi r1Out:
```

3.3.2 Operational Amplifiers

Voltage-controlled amplifiers are commonly implemented using operational amplifiers. *Nodal* provides two operational amplifier models in its component library: a simple, perfect operational amplifier and a more realistic and adaptable model.

The **OpAmp** function assumes that the input resistance is infinite and the output resistance is either 10^{-3} Ω or zero, depending on whether matrix analysis or voltage/current analysis is being undertaken; similarly, depending on the type of analysis, the gain is assumed to be 10^6 or infinite, respectively.

You can use **Draw[OpAmp[]]** to see the circuit that *Nodal* uses:

In:

Draw[OpAmp[]]

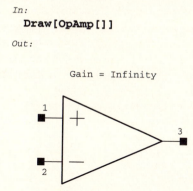

Out:

Gain = Infinity

The simple, perfect model of an operational amplifier is excellent for determining the overall operating characteristics of a circuit.

The **OpAmpModel** function uses a more sophisticated implementation of the operational amplifier. You can specify the input and output resistances, the gain, and the transfer function (which can have multiple frequency knees). You can specify the transfer function using **TransferFunction->1/(1+I*f/fbk)** You must use **f** for the frequency variable and also **fbk** as the frequency breakpoint to allow *Nodal* to equate the variable **fbk** in the transfer function rule to the **fbk** value assigned in the argument list for **OpAmpModel**. **fbk** is the highest frequency at which the open-loop gain of the amplifier can still be obtained; at higher frequencies the maximum realizable gain will be less than the open-loop gain by 20 db per decade of frequency above **fbk**. So, an operational amplifier with an open-loop gain of 100 dB and **fbk**=10 Hz will be able to provide only 80 dB of gain at 100 Hz, 60 dB at 1 kHz, and so on. The following example creates a table of gain as a function of frequency (in the left column):

In:

```
opamp2=NodalNetwork[
            VoltageSource[{1,0},1],
            Resistor[{1,2},1 kOhm],
            Resistor[{2,3},10000 kOhm],
            Resistor[{4,0},1 kOhm],
            OpAmpModel[{4,2,3},
                        1 MOhm,
                        1,
                        10^5,
                        10  ,
            TransferFunction->1/(1+I*f/fbk)]];
NodalAnalyze[opamp2,
            Result->20 Log[10,Abs[V3]],
            Frequency->10^Range[1,10],
            Step->10^Range[1,10]]
```

Out:

Step	$\dfrac{20\ \text{Log}[\text{Abs}[V3]]}{\text{Log}[10]}$
10.	79.1
100.	76.5
1000.	59.9
10000.	40.
100000.	20.
$1.\ 10^6$	-0.0182
$1.\ 10^7$	-20.
$1.\ 10^8$	-40.
$1.\ 10^9$	-60.
$1.\ 10^{10}$	-80.

(1) 1 kΩ (2) 10000 kΩ
OpAmp (3) vOut
(4) +
~ 1 Volt
1 kΩ
(0) Ground

N

3.3.3 Voltage-Controlled Current Sources

Nodal provides both JFET, MOSFET, and MESFET models as well as a pure voltage-controlled current source. Figure 3.6 shows a typical circuit of a simple FET amplifier. In its amplifier mode, the FET is operated in the saturation region and so acts as a voltage-controlled current source.

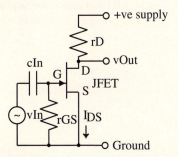

Figure 3.6   FET amplifier.

The voltage across the drain resistor is governed by the channel current, `iDS`, so the gate-source voltage controls the potential at the drain. The ability to change a voltage (at the gate) into a current (in the channel) is called *transconductance*, and is quantitatively identified by the symbol `gm` where

$$\texttt{gm} \;=\; \texttt{gm0}\sqrt{\texttt{iDS/iDSS}}$$

`iDSS` is the drain current with V_{GS}=0 and `gm0` is a constant that depends on the FET's manufactured qualities. `gm0` is always quoted by manufacturers and suppliers on their data sheets as the common-source transconductance. The gain of the amplifier (to a first approximation) is given by the product of `-gm` and `Rd`, where `Rd` is the drain resistor. This equation is straightforward to understand: The value of `gm` indicates how good the FET is at changing a variable voltage at the gate into a variable current, and the voltage across the drain resistor is found from multiplying its resistance by the current through it (Ohm's law). For example, a change of 0.01 V (at I_{DS}=1 mA) will result in a 10 µA change in channel current. For a 10 kΩ drain resistor, a 10 µA current variation will create a fluctuation of 0.1 V: hence, a voltage gain of 10.

You can calculate the drain resistor required to achieve the desired gain by first determining the transconductance, `gm`, at the requested current (`iDSS`=8 mA and `iRequested`=1 mA, say):

```
In:
  gm0=0.01;
  iDSS=0.008;
```

```
iRequested=0.001;
gm=gm0 Sqrt[iRequested/iDSS]
```

Out:
    0.00353553

Using the relation **gain = gm rD**, you can calculate the value of the drain resistor and check the required supply voltage, for the case where the gate-source voltage, **vgs**, is –1.62 V:

In:
```
gain=10;
rD=gain/gm
```

Out:
    2828.43

In:
```
vsupply=iRequested rD - vgs +vds    /.
                              vgs -> -1.62
```

Out:
    4.44843 + vds

The **/.** tells *Mathematica* to apply the rule "**vgs** goes to –1.62" to the preceding equation.

Assuming the FET is biased correctly, *Nodal* provides functions that can help you to look at the AC characteristics of the circuit. If you were to build the FET amplifier in Figure 3.6, and apply power (9 V would be sufficient) and a small AC voltage to the input, you would find that the gain is only about 2 — somewhat errant from your design requirement of 10. Why?

One way to approach this problem is to look at the voltages on all the nodes that might affect the gain. For this example, we have simplified the circuit and use a voltage-controlled current source as the FET. (We have simplified the circuit so as to keep the problem in its most uncluttered form.)

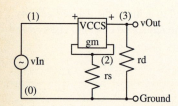

In:
```
Clear[gm];
fet=NodalNetwork[ VoltageSource[{1,0},vIn],
                  VCCS[{1,2,2,3},gm],
                  Resistor[{2,0},rs],
                  Resistor[{0,3},rd]];
NodalAnalyze[fet,
                  Result->{V2,V3}];
Simplify[%]
```

N

```
Out:
  Step              V2                    V3

                    gm rs vIn             gm rd vIn
  1.                ----------            --------------
                    1. + gm rs            -1. - 1. gm rs
```

For the circuit values we have used, the values of **v2** and **v3** can be found:

```
In:
  N[% /. {gm->0.0035,
          rd->rD,
          rs->1.6 kOhm}]

Out:
  Step       V2         V3

  1.         0.85 vIn   -1.5 vIn
```

Remembering that the current through the FET is controlled by the voltage across the gate and source (that is, **vIn–v2**), you are able to see that the source resistor has a drop of **0.85 vIn** across it: The FET sees only (1.00-0.85)**vIn** as its control voltage. The output voltage of **-1.5 vIn** is indeed 10 times the input voltage seen by the FET. Examination of the symbolic equation for **v3** reveals one way around the problem: reduce the small-signal AC impedance of **rs** by bypassing **rs** (whose only purpose is to provide DC bias) at signal frequencies using a capacitor, so preventing the feedback:

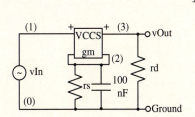

```
In:
  rs=1.6 kOhm;
  rd=2.8 kOhm;
  fetv2=NodalNetwork[VoltageSource[{1,0},vIn],
              Resistor[{2,0},rs],
              Resistor[{0,3},rd],
              VCCS[{1,2,2,3},
                  gm0 Sqrt[iRequested/iDSS]],
              Capacitor[{2,0},100 nF]];

  NodalAnalyze[fetv2,
              Result->V3,
              Frequency->100 kHz]//N

Out:
  (-9.86277 - 0.554611 I) vIn
```

Reanalysis of the circuit with **rs** bypassed will show you that the original design gain of 10 is achieved. But gain is only one of the significant characteristics of a circuit. You can use *Nodal*'s parameter-calculating functions to compute, numerically or symbolically, other characteristics such as input and output impedances, as we show in Chapter 4.

In the analysis of **fetv2**, we specified a frequency at which the analysis was carried out (**Frequency->100 kHz**). Because the internal capacitances of the FET device are small, they did not significantly affect the low-frequency characteristics of the device.

You can repeat the analysis at a number of different frequencies by specifying, for example, **Frequency->{1 kHz, 1 MHz, 1 GHz}**. *Nodal* will step through the analysis three times, using a different frequency for each step.

A list — **{n1, n2, n3,....}** — is one way to enter parameter ranges. *Mathematica* provides other ways of specifying ranges for function arguments.

Toolbox

Creating a Range of Numbers

Range [*max*] generates a list from 1 to *max* in steps of 1; **Range** [*min*, *max*, *delta*] will generate a list from *min* to *max* in steps of *delta*.

Here are four examples of the **Range** function, with the answers given as lists enclosed in **{}**; you can generate both linear and logarithmic series using **Range**.

```
In:
  Range[1,3]

Out:
  {1, 2, 3}

In:
  10^Range[1,3]

Out:
  {10, 100, 1000}

In:
  10^(3 Range[1,3])

Out:
  {1000, 1000000, 1000000000}

In:
  10^Range[1,4,0.2]

Out:
  {10, 15.8489, 25.1189, 39.8107, 63.0957, 100.,
  158.489,251.189,398.107, 630.957, 1000., 1584.89,
```

```
2511.89,3981.07, 6309.57, 10000.}
```
You can use the **Range** function as part of the frequency specification in the function **NodalAnalyze**. To perform circuit analysis every half-decade throughout the range 1 kHz to 1 MHz, you would specify **Frequency -> 10^Range[3,6,0.5]**, for example.

3.3.4 Current-Controlled Current Sources

You can use *Nodal*'s ideal current-controlled current source (**CCCS**) component to establish the fundamental operating characteristics of circuits containing bipolar transistors, because these behave like CCCSs. You should use *Nodal*'s bipolar junction transistor (**BJT**) model for more accurate emulation of real devices.

Note that when you use either the **CCCS** or the **CCVS** component, the controlling current flows from the first to the second node in the component's nodelist: You do not have to provide a path for the controlling current. For example, in **bjtAmp**, **CCCS** provides a short between nodes 2 and 3. The current flowing through this short controls the current that flows from node 4 to node 3:

In:
```
bjtAmp=NodalNetwork[VoltageSource[{1,0},vIn],
                    CCCS[{2,3,3,4},
                         -currentGain],
                    Resistor[{0,4},rc1],
                    Resistor[{3,0},rE1],
                    Resistor[{1,2},rb1]];
bjtGain=NodalAnalyze[bjtAmp,
                     Result->V4]//Simplify
```

Out:
```
      -currentGain rc1 vIn
      --------------------------
      rE1 + currentGain rE1 + rb1
```

3.3.5 Current-Controlled Voltage Sources

Nodal's component library also contains a current-controlled voltage source (CCVS). You can use the **CCVS** for modeling current-differencing (Norton) amplifiers and charge-sensitive amplifiers.

For example, if you need to connect a current source (like a photodiode detector) to an amplifier, does the capacitance of the connecting cable affect the amplifier's output voltage? You can use a current-controlled voltage source to model a charge-sensitive amplifier and a voltage-controlled voltage source to model a voltage amplifier. By specifying **Result->V2**, you can see which amplifier's output voltage does not contain a **cIn** term and is therefore independent of cable capacitance.

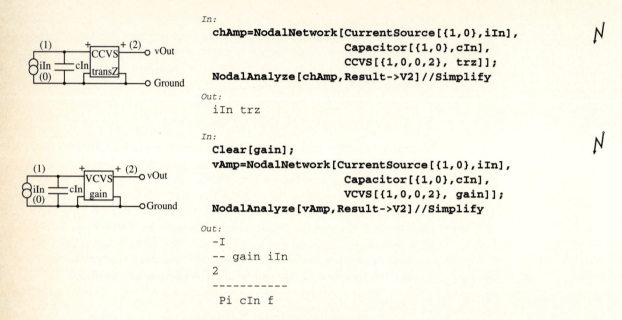

```
In:
  chAmp=NodalNetwork[CurrentSource[{1,0},iIn],
                     Capacitor[{1,0},cIn],
                     CCVS[{1,0,0,2}, trz]];
  NodalAnalyze[chAmp,Result->V2]//Simplify
```

```
Out:
  iIn trz
```

```
In:
  Clear[gain];
  vAmp=NodalNetwork[CurrentSource[{1,0},iIn],
                    Capacitor[{1,0},cIn],
                    VCVS[{1,0,0,2}, gain]];
  NodalAnalyze[vAmp,Result->V2]//Simplify
```

```
Out:
  -I
  -- gain iIn
  2
  -----------
   Pi cIn f
```

You can also use current-controlled voltage sources as convenient ammeters. By passing the current to be measured between the input ports, you can use **NodalAnalyze** to find the voltage between the **CCVS**'s output ports.

3.4 Device-Equivalent Circuits

When you begin to analyze a circuit, it is often useful to keep it as simple as possible so that you can concentrate on the fundamentals of design. Once you have worked out the skeleton of the design, you need to consider smaller side effects, caused by the nonideal way in which all electronic devices behave.

To model FET and BJT devices, *Nodal* uses a subnetwork of components in an equivalent circuit. When you use a FET or BJT device in a numerical calculation, you have to supply values for the resistors and capacitors in the equivalent circuit. In this section, we discuss how you obtain these values from the data supplied by the manufacturer.

3.4.1 The FET Model

The *Nodal* **JFET** component has more details about FET behavior built-in and is therefore more realistic than the simple **VCCS** component. (For example, interelectrode capacitances have to be factored during circuit performance calculations.) You can see *Nodal*'s model of the FET by using the **Draw[JFET[]]** command. The terms in the **JFET** specification refer to the model shown by **Draw** where **Cgs**, **Cgd**, and **Cds** refer to the capacitances between the gate and source, gate and drain, and drain and source, respectively.

In:
> **Draw[JFET[]]**

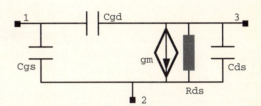

Out:

Sometimes a device's data sheet gives the values of the capacitances **Cgd**, **Cgs**, and **Cds**; **Cds** is very small and can be ignored for most purposes. But often, other names will be used. The most commonly specified capacitances are C_{iss}, the common-source input capacitance, and C_{rss}, the common-source reverse transfer capacitance.

In general, $C_{iss} = C_{gs} + C_{gd}$ and $C_{rss} = C_{gd}$; so $C_{gs} = C_{iss} - C_{rss}$. All capacitances vary with device bias voltages: for n-channel JFETs, they decrease as V_{GS} becomes more negative. C_{iss} ranges from a few pF, for JFETs designed for use in VHF/UHF amplifiers, to a few tens of pF, for JFETs with large internal structures typical of devices used for low on-resistance analog switches.

The transconductance **gm** is a function of the DC conditions and has to be calculated for the particular drain-source current (I_{DS}) used: $g_m = g_{m0}\sqrt{I_{DS}/I_{DSS}}$. For the common-source amplifier configuration, g_{mo} is often specified (at V_{GS}=0) as g_{fs}.

Rds is the dynamic resistance of the drain-source channel and is equal to $1/g_d$, where g_d is the drain conductance or common-source output conductance; $g_d = g_{os} = y_{os} = \partial I_{DS}/\partial V_{DS}$. **Rds** is typically ~$10^3$ to 10^6 Ω and varies strongly with drain current; its value, for your chosen operating point of the FET, should be available from the manufacturer's information.

For a 2N5485 n-channel JFET with I_{DSS} ~ 6 mA, V_{GSoff} ~ –3 V, and g_{mo} ~ 0.004 mho that is operated at V_{DS} = 5 V and I_D = 2 mA, then C_{iss} = 3.5 pF, C_{rss} = 1.5 pF, and r_{ds} ~ 20 kΩ [Siliconix83]. So C_{gs} = 2 pF and C_{gd} = 1.5 pF; C_{ds} is small enough to be ignored for most purposes.

3.4.2 The BJT Model

The model that *Nodal* uses for the bipolar junction transistor, known as the bridged-T model, has the following terms:

In:
Draw[BJT[]]

Out:

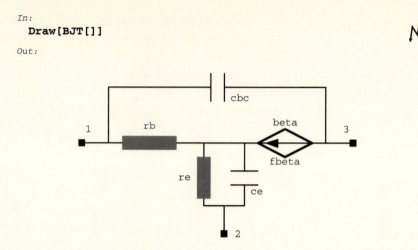

The full *Nodal* BJT function is

BJT[{baseNode,emitterNode,collectorNode},rb,
re,ce,cbc,beta,ftHz,options]

Bipolar transistor parameters are commonly given as *h*-parameters, and so translation of *h* into bridged-T parameters is required. **re** (= r_e and is not to be confused with any device-external resistor connected to the emitter) is given by kT/qI_c or, when I_c is in mA, **re** ~ $25/I_{c(mA)}$ Ω, to a first approximation; **rb** is equivalent to h_{ie} +(h_{fe}+1)r_e. **ce** and **cbc** are the emitter and base-collector capacitances. **beta** is h_{fe} in *h*-parameter models. **fbeta** or **ftHz** is the frequency at which the transistor has unit gain.

You can find values for all these parameters in data sheets — they are mostly functions of other circuit values, such as emitter current (I_E) and the voltage between the emitter and the collector (V_{CE}). For example, a BC109C operated at 2 mA collector current (I_C) and V_{CE}=5 V has h_{fe} ~ 500, h_{ie} ~ 9 kΩ, **ce** ~ 9 pF, **cbc**~ 3 pF, and f_T ~ 160 MHz. Hence, for Nodal's BJT component, **re** ~ 12 Ω and **rb** ~15 kΩ [Philips89].

3.5 Oscillators and Feedback Network Design

Many electronic circuits require feedback from their output to their input. The feedback network — the subcircuit that transmits the fed-back signal — usually must meet a specification that defines the magnitude and phase of the fed-back signal as a function of frequency. You can use *Mathematica* to help you design feedback networks. As our example of this task, we calculate the feedback network for a phase-shift oscillator.

There are two main types of oscillators: relaxation and phase-shift. The typical relaxation oscillator consists of a current source (which can be as crude and compliant as a resistor), a capacitor, and a voltage trigger circuit. The current source charges the capacitor until the capacitor's voltage reaches the trigger circuit's threshold. At that instant, the trigger circuit issues an output pulse and also discharges the capacitor, thus restarting the cycle.

The behavior of the relaxation oscillator is more appropriately studied with time-domain techniques, so we leave further discussion of it until Chapter 7; in the rest of this section, we look at how you can design a phase-shift oscillator with *Mathematica*.

If the output from an amplifier is fed back to its input, the signal will reappear at the output with some shift in phase and change in magnitude. If the phase shift is 2π radians and the product of the amplifier's gain and the attenuation of the feedback loop is at least unity, then the amplifier will oscillate.

The design task for phase-shift oscillators requires you to calculate suitable component values for the phase-shift feedback network (Figure 3.7) so that the network gives the required phase-shift and so that you can determine the minimum amplification factor required to overcome the attenuation of the feedback network (including any effects of the input and output impedances of the amplifier).

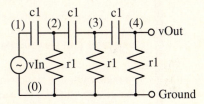

Figure 3.7   Third-order phase-shift network.

As an example, we define a three-stage RC network (see Figure 3.7) to provide π radians (180°) phase-shift. The inverting amplifier provides the remaining π radians of phase-shift to give a total of 2π radians. You specify **Result->V4/V1** to find the ratio of output-to-input voltage:

In:

```
fback=NodalNetwork[VoltageSource[{1,0},vIn],
                   Capacitor[{1,2},c1],
                   Capacitor[{2,3},c1],
                   Capacitor[{3,4},c1],
                   Resistor[{2,0},r1],
                   Resistor[{3,0},r1],
                   Resistor[{4,0},r1]];
```

```
result=NodalAnalyze[fback,
                    Result->V4/V1,
                    Frequency->f]
```

Out:

$$
\frac{8\ I\ Pi^3\ c1^3\ f^3}{8\ I\ Pi^3\ c1^3\ f^3\ -\ r1\ +\ \dfrac{-10\ I\ Pi\ c1\ f}{r1^2}\ +\ \dfrac{24\ Pi^2\ c1^2\ f^2}{r1}}
$$

The **result** is complex because it defines both the magnitude and phase of the output signal relative to the driving input signal. The phase is the argument of this (complex) answer and must be zero when the argument is π radians (that is, the condition for oscillation): You know that, in **result**, the real terms in the denominator must sum to zero so that the remaining complex terms in the denominator cancel the complex numerator. *Mathematica* can do all of the manipulation and so allow you to determine the frequency of oscillation and the minimum amplifier gain to sustain oscillation.

Toolbox

Numerator and Denominator

The numerator and denominator of a fraction can be extracted using the **Numerator** and **Denominator** functions.

First, you create **answer** by removing the list structure from **result**, by using **Normal**, then extract the denominator and the two real terms it contains. You can then determine the frequency of oscillation (at which the real terms are zero), by using **Solve** to determine a rule that defines when the real terms equal zero. Note that you can gather together and select parts of an equation using the same techniques as you would when manipulating a list.

You can find the attenuation of the RC network by calculating the ratio of output-to-input voltage of the network (given by the small-signal analysis) when **f** obeys the rule — **fRule** — found by **Solve**:

In:

```
answer=Normal[result][[1,2]]
```

Out:

$$8\ I\ Pi^3\ c1^3\ f^3$$
$$\overline{\phantom{8\ I\ Pi^3\ c1^3\ f^3\ -\ r1^{-3}\ +\ \cdots\ +\ \cdots\cdots\cdots\cdots\cdots}}$$
$$8\ I\ Pi^3\ c1^3\ f^3\ -\ r1^{-3}\ +\ \frac{-10\ I\ Pi\ c1\ f}{r1^2}\ +\ \frac{24\ Pi^2\ c1^2\ f^2}{r1}$$

In:

denominator=Denominator[answer]

Out:

$$8\ I\ Pi^3\ c1^3\ f^3\ -\ r1^{-3}\ +\ \frac{-10\ I\ Pi\ c1\ f}{r1^2}\ +\ \frac{24\ Pi^2\ c1^2\ f^2}{r1}$$

In:

reals=denominator[[2]]+denominator[[4]]

Out:

$$-r1^{-3}\ +\ \frac{24\ Pi^2\ c1^2\ f^2}{r1}$$

In:

soln=Solve[reals==0, f]

Out:

$$\left\{\left\{f \to \frac{1}{2\ Sqrt[6]\ Pi\ c1\ r1}\right\},\ \left\{f \to \frac{-1}{2\ Sqrt[6]\ Pi\ c1\ r1}\right\}\right\}$$

In:

fRule=soln[[1,1]]

Out:

$$f \to \frac{1}{2\ Sqrt[6]\ Pi\ c1\ r1}$$

In:

attenuation=result /. fRule

Out:

$$-\left(\frac{1}{29}\right)$$

Hence, for the oscillator to work, the gain of the amplifier must overcome the attenuation of the phase-shift network: In this case, the gain must be greater than 29.

You may find the Bode plot useful for a graphical view of phase-shift network output:

In:

```
fback=NodalNetwork[VoltageSource[{1,0},vIn],
                   Capacitor[{1,2},10 nF],
                   Capacitor[{2,3},10 nF],
                   Capacitor[{3,4},10 nF],
                   Resistor[{2,0},6800],
                   Resistor[{3,0},6800],
                   Resistor[{4,0},6800]];
BodePlot[
   NodalAnalyze[fback,
                Result->V4/V1,
                Step->
                Frequency[10^Range[2,4,.05]]]]
```

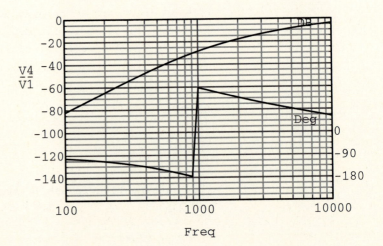

Out:

3.6 Summary

In this chapter, you have seen how

- to calculate the phase and magnitude of voltages and currents in circuits
- to calculate, to plot, and to play the transfer function of RLC circuits
- to compute static and dynamic resistances
- to use simple models to understand the principal concepts of circuits
- to model circuits containing controlled current and voltage sources
- to determine component parameters from manufacturer's data
- to investigate the effect of capacitance on circuit performance
- to investigate and design phase-shift oscillators.

3.7 Exercises

3.1 Write your own functions to convert a complex number into forms that you commonly use. Write a more general form of `complexPolar` (as in Section 3.1.1).

3.2 Investigate the static and dynamic resistance of a JFET for $|Vds|<1$ V.

3.3 Determine the effect of capacitance placed at the output of an amplifier. How are amplifiers affected when constructed with (a) voltage sources and (b) current sources?

3.4 Investigate the effects of amplifier input and output impedances on the operation of a phase-shift oscillator.

3.5 The bandpass of an amplifier driving a capacitive load can be extended by adding an inductor in the output circuit. What is the optimum inductor value?

3.6 Write a *Mathematica* function to calculate the inductor and capacitor values for a tuned-circuit oscillator. What is the Q of this circuit? What are the equivalent values (of R, L, and C) for a crystal?

3.7 Using small-signal models, derive the gain and input and output impedances for (a) common-base, (b) common-emitter, (c) and common-collector transistor amplifiers.

3.8 References

[Philips89] *Data Handbook: Small-Signal Transistors*, Philips Components Ltd., London, 1989.

[Siliconix83] *Small-Signal FET Design Catalog*, Siliconix Inc., Santa Clara, California, 1983.

[Vlach83] Vlach, J., Singhal, K., *Computer Methods for Circuit Analysis and Design*, Van Nostrand Reinhold, New York, 1983.

CHAPTER 4

Small-Signal Circuits II: Multiport Analysis

Until now, we have examined simple circuits with a few components and have calculated the voltage at some node in comparison with the input or ground. However, when we analyze more complicated circuits, we can treat the entire circuit as a black box with an input connection and an output connection. The two ports of the black box — the input and the output — give the analysis method based on this approach its name: two-port analysis. Typically, multiport circuits are analyzed with matrices. Each dimension of the matrix will correspond to a port.

Two-port matrix-based analysis techniques have several advantages:

- They are easily implemented on a computer because the matrices are easily put into forms that relate directly to device measurement or modeling.

- Most manufacturers' data books give device information in either Y-parameters or S-parameters.

- Although you have designed what is inside the blackbox, you do not have to be aware of, say, the voltage at every node when you compute the circuit characteristics.

- The result of a two-port analysis is a simple matrix that summarizes the behavior of the whole circuit.

In this chapter, we examine some of the parameters that you can use to describe circuits, how you go from a netlist to a parameter matrix, how you can solve the parameter matrix, and how you can translate between parameter types.

Matrix analysis methods are fundamental to solving many circuit problems with *Mathematica*, but if you have come to electronics from a scientific — rather than an engineering — background, you may not be familiar with them. Therefore, we shall spend the first section of the chapter reviewing two-port circuits and their matrix descriptions.

4.1 Introduction to the Analysis Method

You can calculate directly different circuit characteristics (such as gain, input and output impedances, and transmitted or reflected power) by using different analysis methods on two-port circuits. The method you choose will depend on what information is available to you and what information you require.

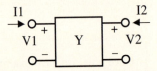

Figure 4.1 Two-port network showing polarity of voltages and currents used in its description.

Figure 4.1 shows a two-port network. Currents and voltages into and out of the network are shown with the subscript 1 associated with the input port and the subscript 2 with the output port. For example, V_2 is the voltage across the two output terminals that constitute the output port.

You can write various equations that describe input or output currents or voltages in terms of voltages or currents at the other terminals and circuit components. The circuit components usually vary with frequency. You can set up the current and voltage equations in many ways, each defining a set of parameters. In all cases, you will use a frequency-domain rather than a time-domain component description. We discuss three commonly used sets of parameters in Sections 4.1.1 through 4.1.3. In Section 4.6, we also discuss the parameters (S-parameters) that are important for high-frequency circuit analysis.

4.1.1 Y-Parameters

Referring to Figure 4.1, you can write the following two equations that describe the interrelation between the currents and voltages in a circuit:

$$I_1 = y_{11} V_1 + y_{12} V_2,$$

$$I_2 = y_{21} V_1 + y_{22} V_2.$$

The quantities y_{11}... are defined and measured when either the input or output port is short-circuited. For example, if the output port is short-circuited, then $V_2 = 0$, since, by definition, a short circuit has (by Ohm's law) no voltage across it. The resulting simple equation gives $y_{11} = I_1/V_1$ when $V_2 = 0$. The constant y_{11} is measured in inverse ohms, or siemens — often called mhos, and is called the input admittance because it is the input current divided by the input voltage.

You can also write these equations in matrix notation as shown below.

$$\begin{bmatrix} I_1 \\ I_2 \end{bmatrix} = \begin{bmatrix} y_{11} & y_{12} \\ y_{21} & y_{22} \end{bmatrix} \bullet \begin{bmatrix} V_1 \\ V_2 \end{bmatrix}.$$

The matrix containing the *y*-constants is called the short-circuit admittance matrix and contains the short-circuit admittance parameters. By convention, these parameters are represented by the symbol *y* and are often referred to as the Y-parameters. Their individual meanings are

y_{11} input admittance, reciprocal of the input impedance

y_{12} reverse transfer admittance

y_{21} forward transfer admittance

y_{22} output admittance, reciprocal of the output impedance.

All the Y-parameters are measured in siemens, with the input and forward parameters measured with $V_2 = 0$ and the output and reverse parameters measured with $V_1 = 0$.

4.1.2 Z-Parameters

Similarly, you can write equations that describe the input and output voltages in terms of the input and output currents and constants, named the Z-parameters, which describe various impedances associated with the circuit. The Z-parameters are measured in ohms and are defined with input and output currents zero, so they represent open-circuit conditions. Using matrix notation, you can write

$$\begin{bmatrix} V_1 \\ V_2 \end{bmatrix} = \begin{bmatrix} z_{11} & z_{12} \\ z_{21} & z_{22} \end{bmatrix} \bullet \begin{bmatrix} I_1 \\ I_2 \end{bmatrix}.$$

You can solve for any *z* by making either I_1 or I_2 zero. For example, $z_{11} = V_1/I_1$ when $I_2 = 0$.

The individual meanings of the Z-parameters are

z_{11} input impedance

z_{12} reverse transfer impedance

z_{21} forward transfer impedance

z_{22} output impedance.

4.1.3 ABCD-Parameters

You can reorder the basic matrix notation in a number of ways to adapt it to a particular purpose. When you are synthesizing a larger system from its subunits, you may want to chain together circuit modules. You can compute the behavior of the whole system simply by multiplying together the *ABCD*-matrix description of each subunit. You can analyze any chain of components by multiplying together the *ABCD*-matrices of the individual elements to get the *ABCD*-matrix of the overall chain. *ABCD*-matrices are popular for analyzing filters because most filter networks form a ladder, or cascade, of series and shunt elements.

The term *chain* derives from the German literature, which refers to this type of matrix as *Kettenmatrix* or "chain matrix" in English [Pohl67]. The term *ABCD* arose from the convention of calling the four elements in this two-port matrix *A*, *B*, *C*, and *D*, respectively.

So that chaining is possible, the terminal voltages and currents must be in the same vector. Both *Y*-parameters and *Z*-parameters have voltages in one vector and currents in the other. The equations for ABCD-parameters are, in matrix form:

$$
\begin{bmatrix} V_1 \\ I_1 \end{bmatrix} = \begin{bmatrix} A & B \\ C & D \end{bmatrix} \bullet \begin{bmatrix} V_2 \\ -I_2 \end{bmatrix} = \begin{bmatrix} a_{11} & a_{12} \\ a_{21} & a_{22} \end{bmatrix} \bullet \begin{bmatrix} V_2 \\ -I_2 \end{bmatrix} .
$$

The a_{ij} have the following meanings:

a_{11} reciprocal of the open-circuit forward voltage gain

a_{12} reciprocal of the short-circuit forward transfer admittance

a_{21} reciprocal of the open-circuit forward transfer impedance

a_{22} reciprocal of the short-circuit forward current gain.

4.1.4 Parameter Calculation

In Sections 4.1.1 through 4.1.3 we reviewed some of the two-port parameters and, beginning with Section 4.2, we will show you, in some detail, how *Mathematica* can be programmed to carry out the calculation of parameters. In this section, we introduce you to the functions within *Nodal* that carry out parameter calculation, just in case you do not want to wait for the methodological discussion.

Nodal provides you with two ways of calculating parameters. You can specify the parameters you want in the **NodalAnalyze** function, or you can use the parameter-dedicated functions **ABCDParameters**, **SParameters**, **TParameters**, **YParameters**, and **ZParameters**.

To use the `NodalAnalyze` function, you state which parameter or parameters you require as part of the `Results` option. For example, you can calculate, with one call to `NodalAnalyze`, the input and output impedances (`Z11` and `Z22`), the voltage at node 2 (`V2`), and the forward transfer function of a circuit (`A11`):

In:

```
myNet=NodalNetwork[VoltageSource[{1,0},vIn],
                   Resistor[{1,2},10],
                   Resistor[{2,0},20],
                   Resistor[{2,3},30],
                   Resistor[{3,0},40],
                   Resistor[{3,4},50]];
NodalAnalyze[myNet,Result->{Z11,V2,A11,Z22},
    Nodes->{1,4}]
```

Out:

Step	Z11	V2	A11	Z22
1.	25.6	0.609 vIn	2.88	72.2

The `Nodes` option is new, and it is essential that you use it to specify which nodes are input and which nodes are output ports. You can have many nodes and many ports in your circuits, but the rest of this section will assume that you have only two ports. Complicated circuits have many nodes, so you have to tell *Nodal* which nodes are the input and which nodes are output ports. The `Nodes` option requires a two-element list (or longer for multiports) containing the node number of the input and output nodes, both of which are referenced to the ground node, node 0. So `Nodes->{1,4}` tells *Nodal* that the input signal is connected between node 1 and ground (node 0) and that the output signal is between node 4 and ground. *Nodal* automatically maps the node numbers in the `Nodes` option onto the conventional notation for input and output nodes in two-port circuits. For example, `Result->Z11` will calculate the input impedance for a signal applied between node 1 and ground, providing that you have specified, in the `Nodes` option, that node 1 is the input node.

The `Step` number refers to the number of analyses performed. If we had requested analysis, at, say, three different frequencies, then the output would have occupied three rows, one for each frequency, and each row would have been labeled with a different `Step` identifier. By default, *Nodal* numbers the steps, starting with 1, but you can specify the step identifiers by placing them in a list as the argument for the `Step` option:

```
In:
    myNet2=NodalNetwork[VoltageSource[{1,0},vIn],
                        Resistor[{1,2},10],
                        Resistor[{2,0},20],
                        Resistor[{2,3},30],
                        Capacitor[{3,0},47 nF],
                        Resistor[{3,4},50]];
    NodalAnalyze[myNet2,Result->{Z11,DB[V2/V1]},
    Nodes->{1,4},Frequency->{10,100,1000}]
```

Out:

Step	Z11	$DB\left[\dfrac{V2}{V1}\right]$
1.	30. - 0.00118 I	-3.52
2.	30. - 0.0118 I	-3.52
3.	30. - 0.118 I	-3.52

```
In:
    NodalAnalyze[myNet2,Result->{Z11,DB[V2/V1]},
    Nodes->{1,4},Frequency->{10,100,1000},
    Step->{10,100,1000}]
```

Out:

Step	Z11	$DB\left[\dfrac{V2}{V1}\right]$
10.	30. - 0.00118 I	-3.52
100.	30. - 0.0118 I	-3.52
1000.	30. - 0.118 I	-3.52

If you want one parameter type calculated, then you can use one of the dedicated functions. For example, to calculate the admittance matrix of a circuit, you use **YParameters**:

```
In:
    N[YParameters[myNet2,
                  Nodes->{1,4},
                  Frequency->100 kHz],3]
```

```
Out:
  0.0404 + 0.00314 I      -0.00553 + 0.00346 I

 -0.00553 + 0.00346 I    0.0139 + 0.0038 I
```

Although we have used a numeric example, you can obtain a symbolic answer if some of the circuit components have no numeric value or if you do not give numeric arguments to the **Frequency** option. Symbolic answers can be very long.

Should you want to convert the Y-parameter result into another parameter (Z- or S-parameter, for example), then we show you how in Section 4.4.

4.2 Matrix Analysis with *Mathematica*

A general circuit analysis package works by manipulating matrices (as described in Section 4.1) and so requires a matrix as input. However, the starting point for an engineer is a netlist that lists the circuit's components and the nodes between which they are connected. To apply matrix methods, we must first create a program to translate the netlist into a matrix for analysis. You can find more about the relationships among meshes, nodes, and matrices in Vlach [Vlach83].

We begin by converting a netlist into Y-parameters, because it is easily programmed. Rather than adopt a formal approach, we provide a simple example and a few observations to help you to understand the algorithm that creates a nodal circuit matrix from a netlist.

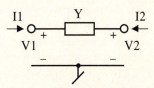

Figure 4.2   Two-node series circuit used to demonstrate algorithm for converting a netlist into Y-parameters.

The circuit of Figure 4.2 forms a basic two-node component and shows how any component between two nodes is put into a Y-matrix. This method is general because all components will be between two nodes or between a node and ground. Taking the two-node circuit shown in Figure 4.2, we can write the equations

$$I_1 = Y(V_1 - V_2),$$
$$I_2 = Y(V_2 - V_1).$$

Once we represent this trivial set of equations as a matrix, the general algorithm for nodal-based CAD becomes obvious:

$$\begin{bmatrix} I_1 \\ I_2 \end{bmatrix} = \begin{bmatrix} Y & -Y \\ -Y & Y \end{bmatrix} \bullet \begin{bmatrix} V_1 \\ V_2 \end{bmatrix} \;.$$

In the matrix representation of this component, each place on the matrix diagonal contains the component admittance, Y, whereas each off-diagonal value is the negative of the admittance. Although we have used only a simple two-node circuit, the result is general: Matrix elements y_{ii} are loaded with Y, and matrix elements y_{ij} ($i \neq j$) are loaded with $-Y$. Note that each row and column of this matrix has values that sum to zero. The zero sum serves as a useful check and a reminder of Kirchhoff's law. The matrix is called an indefinite admittance matrix because we have not defined the ground node. The indefinite admittance matrix is singular, and so cannot be inverted because no ground reference node is established. Once we define a ground node, we set its voltage to zero, canceling the row and column in the matrix that correspond to the ground node. The new matrix is called a definite admittance matrix and can be inverted to solve for the node voltages.

Now we have a simple algorithm for nodal-based CAD: Components between two nodes have their admittance added to the diagonal at a position representing each node and have the negative of their value added at the off-diagonal position representing each node pair. The ground node is usually set to the highest row and column of the matrix. The last row and column of the matrix are eliminated before inverting, which corresponds to changing the indefinite to a definite matrix by establishing a ground node.

Our matrix representation needs two vectors: a vector of node voltages for the values of each node voltage referenced to ground and a vector of the currents into each node. Because it would be difficult to know all node voltages, and it is easy to know the currents applied to each node, the nodal description is especially convenient. You know the node voltages once the Y-matrix is inverted and then multiplied by the current vector.

The use of component admittances does create one complication in that ideal voltage sources cannot be used, because their zero impedance would create an infinite admittance. For now, we shall require all voltage sources to have at least a small finite resistance.

A small resistance allows voltage sources to be transformed, through a Norton equivalence, to a current source with a parallel resistance. Figure 4.3 shows the Norton equivalent current source for a voltage source and series resistor. The important criterion for these equivalent forms is that the circuits be indistinguishable at the terminals. If each circuit has the same open-circuit voltage and short-circuit current, then these circuits are equivalent. The short-circuit current is **Vs/R1** and the open-circuit voltage is **Vs**. (When a current source and shunt resistor are transformed into an equivalent voltage source and series resistor, respectively, the latter is known as a Thévenin equivalent circuit.

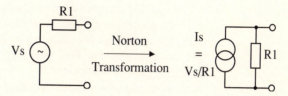

Figure 4.3 Norton equivalent of a voltage source and resistor in series.

Figure 4.4 shows a third-order network with a Norton-transformed source — to make the network suitable for nodal analysis. The following analysis uses the network of Figure 4.4 for clarity.

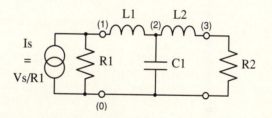

Figure 4.4 Third-order circuit suitable for nodal analysis.

We use Y-parameters for CAD because we can define the *Y*-matrix by inspecting the circuit and we need to know only those currents applied to the circuit from external sources. An example will help us to consolidate all this. For a four-node circuit, we first need to define a 4×4 matrix.

```
In:
   yMatrix = DiagonalMatrix[Table[0,{4}]];
```

Because we initially use an indefinite matrix, we define the ground node to be the fourth node. Using the algorithm just described, we can load the first component — a resistor, **R1** — which we connect between nodes 1 and 4, into our matrix. We first load the component's admittance (1/**R1**) into elements y_{ii} (i = first node, second node) and then load the negative of the admittance into elements y_{ij} (i,j = first node, second node; $i \neq j$).

```
In:
    Apply[(yMatrix[[##]] += 1/R1)&,
          {{1,1},{4,4}},
          1];
```

We fill the diagonal of **yMatrix** by using a list of indices, supplied as the second argument to the **Apply** function. The **##** designator declares that the entire list is to be used — element by element — as the full set of arguments. This is equivalent to setting **yMatrix[[1,1]] += 1/R1** and **yMatrix[[4,4]] += 1/R1**.

The **&** after the **(yMatrix[[##]] += 1/R1)** is a postfix form of **Function** and instructs *Mathematica* to treat the command's argument as a function with its own arguments, which are declared using the **#** or **##** notation. **(yMatrix[[##]] += 1/R1)** is an *anonymous function*.

Toolbox

Declaring Pure (Anonymous) Functions

You can define functions using the form **fnName[arg_]:=** *fnBody*, where *fnBody* represents the *Mathematica* statements for that function. However, *Mathematica* allows a shorter version of function definition that does not employ explicit names for arguments or for the function. Functions defined in this way are called pure, or anonymous, functions. You may find pure functions useful where a user-defined function is helpful but will not be used elsewhere.

The analogous definition of **fnName** in the shorthand form of a pure function is **(fnBody)&**, where the **&** symbol following the definition identifies the preceding code as a pure function. The position of the arguments supplied to the pure function is indicated by the **#** symbol.

The definition of pure functions is an important topic, and we give more information in the Appendix; we also recommend that you read more about it in Stephen Wolfram's book *Mathematica — A System for Doing Mathematics by Computer* [Wolfram91].

You can then display the matrix to see the results of the first part of the loading process:

```
In:
  yMatrix //MatrixForm
```
```
Out:
    1
    --    0    0    0
    R1

    0     0    0    0

    0     0    0    0

                    1
    0     0    0    --
                    R1
```

The off-diagonal components are then filled in a similar fashion:

```
In:
  Apply[(yMatrix[[##]] += -1/R1)&,
        {{1,4},{4,1}},1];
  yMatrix //MatrixForm
```
```
Out:
    1                      1
    --      0      0     -(--)
    R1                     R1

    0       0      0       0

    0       0      0       0

    1                      1
  -(--)     0      0      --
    R1                     R1
```

Although this method of loading the matrix is compact and fast, it is better to have functions that convert circuit elements into a submatrix which a **Do** loop can then load into the final matrix. In general, **Do** loops are slow and do not make efficient use of *Mathematica*, but sometimes they provide clear and effective programming. We will use a **Do** loop to insert each submatrix into the larger matrix, **yMatrix**. A submatrix function, **Yelement**, creates a submatrix from each admittance. The **Do** loop is set up for a 2×2 matrix and uses the list of component nodes to determine where in the *Y*-matrix the component belongs. You can easily extend this method to a circuit analysis program because it is easy to extract the nodes from a netlist and to convert the element names into functions that return a submatrix. For example, you can use this new method to load an inductor, **L1**, between nodes 1 and 2 of a circuit:

```
In:
  Yelement[x_]:= {{x,-x},{-x,x}}
  subMatrix = Yelement[1/(p L1)];
  nodes = {1,2};
  Do[yMatrix[[nodes[[i]], nodes[[j]] ]] +=
                            subMatrix[[i,j]],
        {i,2}, {j,2}];
  yMatrix //MatrixForm

Out:
    1     1          1                    1
    -- + ----     - (----)        0     - (--)
    R1    L1 p       L1 p                  R1


      1            1
  - (----)        ----           0        0
      L1 p         L1 p


    0            0               0        0


    1                                     1
  - (--)          0               0       --
    R1                                    R1
```

You can see how **Yelement**'s action can be automated because only the element values and the node numbers are required. You can load an arbitrary circuit element into the matrix given only the list of nodes and the component's admittance. Loading would be done for each element in a netlist, and each line of the netlist (network) would contain the admittance and nodes of the element. We show an automated version in the following example.

We used the following network to demonstrate mesh analysis in Section 3.1.2. In this chapter, we will derive the same transfer function as in Section 3.1.2 by nodal analysis and by *ABCD*-matrix cascade analysis. You can compare the complexity and ease of automation of each approach.

We use **Apply** to map **Yelement** over the third term in each row of the netlist named **network**, and we extract the nodes into a list in a similar manner. Finally, the **Do** loop that loads the matrix is mapped over the lists.

Extract nodes from network list.

Convert network listing to matrices.

```
In:
  network = {{2,3,1/(p L2)},
              {2,4,p C1},
              {3,4,1/R2}};
  subMatrices = Apply[Yelement[#3]&,network,1];
  nodes = Apply[{#1,#2}&,network,1];
```

Load main matrix from extracted nodes and matrices.

```
Do[yMatrix[[ nodes[[#]][[i]], nodes[[#]][[j]] ]]
              += subMatrices[[#]][[i,j]],
    {i,2},
    {j,2}
  ]& /@ Range[Length[network]];
```

```
yMatrix //MatrixForm
```

Out:

$$
\begin{pmatrix}
\dfrac{1}{R1} + \dfrac{1}{L1\ p} & -\left(\dfrac{1}{L1\ p}\right) & 0 & -\left(\dfrac{1}{R1}\right) \\[2ex]
-\left(\dfrac{1}{L1\ p}\right) & \dfrac{1}{L1\ p} + \dfrac{1}{L2\ p} + C1\ p & -\left(\dfrac{1}{L2\ p}\right) & -(C1\ p) \\[2ex]
0 & -\left(\dfrac{1}{L2\ p}\right) & \dfrac{1}{R2} + \dfrac{1}{L2\ p} & -\left(\dfrac{1}{R2}\right) \\[2ex]
-\left(\dfrac{1}{R1}\right) & -(C1\ p) & -\left(\dfrac{1}{R2}\right) & \dfrac{1}{R1} + \dfrac{1}{R2} + C1\ p
\end{pmatrix}
$$

The preceding code forms the basis for a simple analysis program. Note that each row and column in the matrix sum to zero, which is a property of the indefinite admittance matrix. Next, we must convert the matrix to a definite Y-matrix by removing the last row and column. The following code takes the first three elements of the first three rows and sets the matrix to this reduced matrix, a method that is simpler and more readable than transposing and dropping rows and columns:

In:
```
yMatrix = yMatrix[[ Range[3],Range[3] ]];
```

We next define an input current vector based on the Norton equivalent of a voltage source. Our solution for V3 is

In:
```
currentVector = {Vs/R1, 0, 0};
voltageVector = Inverse[yMatrix].currentVector;
voltageVector[[3]]
```

```
Out:
                         2   C1        C1
        Vs / (L1 L2 R1 p  (----- + ----- +
                             L2 R1    L1 R2


               1                 1
        ----------- + ----------- +
               2                 2
        L1 L2 R1 p     L1 L2 R2 p


          C1            1             1          C1 p
        ------- + ---------- + ---------- + -----))
        L1 L2 p    L1 R1 R2 p    L2 R1 R2 p    R1 R2
```

Although the result looks unfamiliar at first, we can convert it to a transfer function and then simplify it to get the **transferFunction** of Section 3.1.2:

```
In:
   nodeMatrixFunction =
                Simplify[ voltageVector[[3]]/Vs]
Out:
                                                                    2
        R2/(R1 + R2 + L1 p + L2 p + C1 R1 R2 p + C1 L2 R1 p

                            2              3
               + C1 L1 R2 p  + C1 L1 L2 p  )
```

4.3 General Matrix Analysis

In Section 4.1, we described a few of the matrix equations that describe two-port circuit characteristics. Each form of matrix has different advantages in measurement or manipulation, depending on what information you wish to determine and what information you have available to use in the solution process. Once we have defined a circuit in one set of parameters, we often must evaluate circuit performance via another type of matrix equation, so being able to convert between one representation and another is essential.

The simplest transformation is between a Y-matrix and a Z-matrix. Inverting a definite Y-matrix converts it to an impedance (Z) matrix. By rearranging the voltage and current vectors, we can derive several other useful matrices, the most important being ABCD- and S-parameter matrices. In Section 4.4, we present a tour of some of these matrix transformations.

4.3.1 *ABCD*-Matrices

We derive *ABCD*-matrices for any two-port circuit by applying the following equation to series and shunt network elements. Note that *ABCD*-matrices use the voltage and current directions shown in Figure 4.5; *Z*-, *Y*-, *G*-, and *H*-matrices use the directions given for the *Y*-matrix also shown in Figure 4.5.

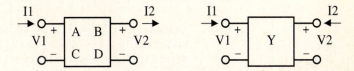

Figure 4.5   Voltage and current directions for an *ABCD*-matrix and a *Y*-matrix.

As a reminder, here is the *ABCD*-matrix equation:

$$\begin{bmatrix} V_1 \\ I_1 \end{bmatrix} = \begin{bmatrix} A & B \\ C & D \end{bmatrix} \cdot \begin{bmatrix} V_2 \\ -I_2 \end{bmatrix}.$$

A series element, **z**, and a shunt element, **y**, have the *ABCD*-matrices

```
In:
    {{1,Z},{0,1}} //MatrixForm
Out:
   1  Z
   0  1


In:
    {{1,0},{Y,1}} //MatrixForm
Out:
   1  0
   Y  1
```

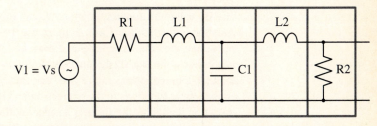

Figure 4.6   Five-component filter network arranged as a chain of elements.

All the circuit elements used in the example in Section 4.2 are either series or shunt elements and form a typical ladder network. By implementing a few functions to cascade the elements shown in Figure 4.6, we get a basic ladder network analysis program. The first two functions each convert a series or shunt element value to an *ABCD*-matrix. The multiplication of the individual element matrices gives the transfer function matrix, **tfMatrix**, of the whole cascaded circuit:

In:

```
ABCDSeries[z_] := {{1,z},{0,1}}
ABCDShunt[y_] := {{1,0},{y,1}}
tfMatrix = ABCDSeries[R1].
    ABCDSeries[p L1].ABCDShunt[p C1].
        ABCDSeries[p L2].ABCDShunt[1/R2];
```

Note that *Mathematica* uses a '.' (dot product) for matrix multiplication. For clarity, the sequence of matrix multiplications is shown here:

$$\begin{bmatrix} A & B \\ C & D \end{bmatrix} = \begin{bmatrix} 1 & R1 \\ 0 & 1 \end{bmatrix} \bullet \begin{bmatrix} 1 & L1p \\ 0 & 1 \end{bmatrix} \bullet \begin{bmatrix} 1 & 0 \\ C1p & 1 \end{bmatrix} \bullet \begin{bmatrix} 1 & L2p \\ 0 & 1 \end{bmatrix} \bullet \begin{bmatrix} 1 & 0 \\ \frac{1}{R2} & 1 \end{bmatrix} \quad .$$

The matrix equation leads to complicated terms in the final matrix. Our result is simplified by realizing that only the **[[1,1]]** term, the *A*- term, is needed. In our example, $I_2 = 0$ and the *A*-element of the complete chain gives V_1/V_2 when I_2 is zero: We obtain the transfer function (V_2/V_1) by inverting the *A*-term:

In:

```
tfABCD = Simplify[ 1/Expand[tfMatrix[[1,1]]] ]
```

Out:

$$R2 \; / \; (R1 + R2 + L1 \; p + L2 \; p + C1 \; R1 \; R2 \; p + C1L2 \; R1 \; p^2 +$$

$$C1 \; L1 \; R2 \; p^2 + C1 \; L1 \; L2 \; p^3 \;)$$

By using the **Expand** function, we can force **Simplify** to put the equation into a familiar form. You can use the *Mathematica* functions **Together**, **Expand**, **FactorSquareFree**, and **Cancel** to simplify expressions; the **Simplify** function in *Mathematica* version 2.2 uses **Expand**, **ExpandDenominator**, and **FactorSquareFree**. **Simplify** applies these functions to the expression and its subexpressions, and uses the operation that gives the shortest answer [Withoff91]. When you work with trigonometric expressions, remember that **Simplify** uses **Trig->True** as a default, while **Expand** and **FactorSquareFree** use **Trig->False** as a default.

By comparing the last result against the **transferFunction** of Section 3.1.2, we can reassure ourselves that no mistakes have been made:

In:
```
    tfABCD == transferFunction
```

Out:
```
    True
```

4.4 Matrix Transformations

The next two equations define the transformation from Y- to ABCD-parameters. Different current directions must be accounted for in the transformation because the direction of I_2 in for *ABCD*-matrices is opposite to the definition used in *Y*-matrices; the definition used by the ABCD-parameters allows the chain multiplication to work, but it also causes confusion.

To equate the two matrix forms we take the negative sign from I_2 and move it to the *B* and *D* matrix terms. Rearranging the *ABCD*-matrix equation and then using **Solve** gives the *Y*-matrix elements in terms of the *ABCD*-matrix elements:

In:
```
    ySoln =
    Solve[{V1,I1} == {{A,-B},{C,-D}}.
                        {V2,I2},{I1,I2}]
```

Out:
```
                      D (V1 - A V2)
    {{I1 -> C V2 + -------------,
                          B

                      V1 - A V2
          I2 -> -(--------)}}
                        B
```

We find both y_{11} and y_{21} by setting V_2 to zero, equivalent to placing a short circuit on port 2. We calculate y_{12} and y_{22} with a short circuit placed on port 1, causing V_1 to be zero. By substituting the known currents into the respective current-to-voltage ratio and setting the proper port voltage to zero, we get our solution:

In:
```
    ({y11 -> I1/V1, y21 -> I2/V1} /. ySoln) /. V2->0
```

Out:
```
                D            1
    {{y11 -> -, y21 -> -(-)}}
                B            B
```

```
In:
   Simplify[({y12 -> I1/V2, y22 -> I2/V2} /. ySoln)
           /. V1->0]
```

```
Out:
                  A D           A
   {{y12 -> C - ---, y22 -> -}}
                  B             B
```

Each matrix transformation problem is solved in the above manner. For 2×2 circuits, solutions can be found in various textbooks [Ghausi65, Kuo75]. For the engineer, the main issue is to have the transformations readily accessible. We think the functional programming form in *Mathematica* is especially helpful because it allows us to use the transformed results in further analysis and in larger programs so that we can build on our earlier work. *Mathematica*'s extensibility contrasts with dedicated analysis programs, which may require you to translate the format of results before the results can be used by the next dedicated program.

It is tempting to create functions for each matrix transformation and to give them names such as **ABCDtoY**. A plethora of functions requires us to remember many function names and forces us to remember what parameter type each data set represents. A better approach uses object-oriented techniques to organize transformation functions.

4.4.1 Object-Oriented Technique

If we tag each object of data and have our functions recognize those tags, then we can rely less on our memory and more on smart programming. Because *Mathematica* allows us to wrap any name we wish around a set of data, we can set up functions that react differently to the different types of data passed to them. For example, we might wrap the matrix {{1,p L},{0,1}} with the name **ABCDData**, thus forming an *ABCD*-matrix object. We then have a set of functions such as **YParams**, **ZParams**, **ABCDParams**, and **SParams** that automatically convert each object appropriately. *Nodal* contains more general forms of all the preceding functions. However, *Nodal* uses full names, such as **YParameters**, for functions. Also, *Nodal* uses **ABCDMatrix** to wrap *ABCD*-data. We use different names here to avoid conflicts with *Nodal*.

The last results in the introduction to Section 4.4 form the following transformation which gives the *Y*-matrix of a series element.

```
In:
   YParams[ ABCDData[{{a_,b_},{c_,d_}}] ] :=
              YData[ {{d/b,c - a d/b},
                       {-1/b,a/b}} ];

   YParams[ ABCDData[{{1,p L},{0,1}}] ]
```

Out:

$$YData[\{\{\frac{1}{L\,p},\ -(\frac{1}{L\,p})\},\ \{-(\frac{1}{L\,p}),\ \frac{1}{L\,p}\}\}]$$

A secondary advantage to using objects is that *Mathematica*'s **Format** command can recognize each object and display that object in an optimal way. For example, scattering matrices need to carry information about the port impedance normalization, yet you do not always need to see the normalization values. So we use **Format** to show only the matrix:

In:
```
Format[ YData[a_] ] := MatrixForm[a]
Format[ ABCDData[a_] ] := MatrixForm[a]
```

Using this new format gives you a more readable result, as we show by redoing the transfer function example from Section 3.1.2:

In:
```
YParams[ ABCDData[{{1,p L},{0,1}}] ]
```

Out:

$$\begin{matrix} \dfrac{1}{L\,p} & -(\dfrac{1}{L\,p}) \\[2ex] -(\dfrac{1}{L\,p}) & \dfrac{1}{L\,p} \end{matrix}$$

4.5 CAD Program

Using the ideas in Section 4.4, we can construct a small CAD program. We shall use the same nomenclature of **YParams** for the transforming function and **YData** for the data type, and we shall extend the function **YParams** to recognize the data type **Circuit**. Then we shall build an analysis function called **SolveCircuit** to solve for the node voltages. This allows us to solve easily the earlier example problem, as well as many other problems.

First, we define a basic netlist that identifies the circuit's components and their node connections:

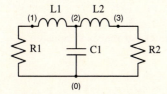

In:
```
myCirc = Circuit[Conductance[{1,0},1/R1]
                 Conductance[{1,2},1/(p L1)],
                 Conductance[{2,0},p C1],
                 Conductance[{2,3},1/(p L2)],
                 Conductance[{3,0},1/R2]];
```

Note that we never define **Circuit** — it serves only as a wrapper to identify the data. The variable **myCirc** now holds the netlist. Here, we use node 0 as ground, and, for simplicity, we use only one type of component, a **Conductance**. **Conductance** is also a wrapper.

Next, we define a new **YParams** function. **YParams** mainly uses the code defined in Section 4.1.1. Because the functions in Section 4.1.1 worked over lists, we must remove the **Circuit** wrapper and define **network** to be the list of the components:

```
In:
  network = Apply[List, myCirc]
```

```
Out:
                             1                              1
  {Conductance[{1, 0},  --],  Conductance[{1, 2},  ----],
                            R1                            L1 p
                                                           1
  Conductance[{2, 0},  C1 p],  Conductance[{2, 3},  ----],
                                                          L2 p

                            1
  Conductance[{3, 0},  --]}
                           R2
```

Then we extract the nodes as we did before. This time the first argument to **Conductance** is a list containing both nodes, so the extraction is easy:

```
In:
  nodes = Apply[#1&,network,1]
```

```
Out:
  {{1, 0}, {1, 2}, {2, 0}, {2, 3}, {3, 0}}
```

We shall use these nodes in a matrix, so the zero node will create a problem because *Mathematica* numbers the elements of an n-long list 1 to n (as does FORTRAN) rather than 0 to $n-1$ (as does C). We must therefore offset all nodes by 1 so that the ground will appear in the first row and column. This method is easier to automate than one that has the ground node as the highest node.

```
In:
  nodes += 1
```

```
Out:
  {{2, 1}, {2, 3}, {3, 1}, {3, 4}, {4, 1}}
```

We now find the highest node, because it determines our matrix dimension:

In:
```
maxNode = Max[nodes]
```

Out:
```
4
```

We can now build our revised **Yelement** function. The *Y*-matrix is constructed as before, with the **Yelement** function being used to turn the components into submatrices. The last line converts the indefinite matrix into a definite one by using all the rows and columns but the first. The end result is that **YParams** is able to recognize a netlist as well as other matrix types.

In:
```
YParams[ circ_Circuit ] :=
 Module[ {subMatrices,nodes,network,
                          yMat,maxNode},
   network = Apply[List,myCirc]; (* to List *)
   nodes = 1 + Apply[#1&,network,1];(*+offset*)
   maxNode = Max[nodes];
   subMatrices = Apply[Yelement[#2]&,network,1];
   yMat = DiagonalMatrix[Table[0,{maxNode}]];
        (* fill the matrix *)
   Do[
     yMat[[ nodes[[#]][[i]],
        nodes[[#]][[j]] ]] +=
                 subMatrices[[#]][[i,j]],
                {i,2},{j,2}]& /@
                    Range[Length[network]];
        (* convert to definite, remove node 1 *)
   YData[yMat[[Range[2,maxNode],
                     Range[2,maxNode]]]]
 ]
```

Convert Circuit wrapper to List.
Extract nodes from netlist.
Create submatrices from netlist.
Define an empty matrix based on nodes required.

Fill the matrices via the submatrix list and nodes list.

Convert final matrix to definite Y by removing ground node.

The **YData** wrapper allows the formatting form, discussed in Section 4.4.1, to work with the result of **YParams**:

In:
```
YParams[ myCirc ]
```

Out:
```
   1     1                    1
   -- + ----                  -(----)                          0
   R1    L1 p                  L1 p


       1                    1     1                    1
   -(----)                  ---- + ---- + C1 p     -(----)
       L1 p                 L1 p   L2 p               L2 p


                                 1                    1     1
       0                     -(----)                  -- + ----
                                 L2 p                 R2    L2 p
```

For completeness, we include a help message. We also give the user a way to turn the **YData** object into a list. A list is easily understood by native *Mathematica* functions. The standard way to do this is to make the **Normal** function change the head from **YData** to **List**:

In:
```
YParams::usage=
"YParams[ABCDData[..] or Circuit[..]]
returns the Y parameter matrix (as YData[..]) for
the given matrix data, such as ABCDData, or for
the given Circuit data. YParams works like
YParameters in Nodal but a different name was used
to avoid conflicts.";
```

In:
```
Normal[YData[x_]] ^= x
```

Out:
```
x
```

Finally, we create a function for solving our original circuit. The equations in Section 4.2 are put into the following function:

In:
```
SolveCircuit[YData[y_], currents_List]:=
                          Inverse[y].currents
```

Now, our original problem is solved as shown here:

In:
```
volts = SolveCircuit[ YParams[myCirc],
                     {Vs/R1,0,0}];

voltageTF = Simplify[ volts[[3]]/Vs]
```

Out:

$$R2/(R1 + R2 + L1\ p + L2\ p + C1\ R1\ R2\ p + C1\ L2\ R1\ p^2 +$$

$$C1\ L1\ R2\ p^2 + C1\ L1\ L2\ p^3)$$

Once again, we confirm the equality of our results:

In:
```
voltageTF == transferFunction
```

Out:
```
True
```

4.6 S-Parameters

Our final discussion on matrices in this chapter concerns the scattering parameters, or S-parameters. S-parameters differ from other matrix descriptions for many reasons. One of the most significant reasons is that S-parameters assume that all ports (external nodes) are terminated. Terminating ports makes S-parameters the parameters of choice for higher-frequency work, as we will discuss in Chapter 9.

For example, analysis of the circuit in Section 4.4 would include `R1` and `R2` automatically; the circuit description would need to specify only the inductors and the capacitor. To support S-parameters we could use the function `SolveCircuit` with an option, `Ports->R1`, to define the port impedances.

S-parameters have gained in popularity since their introduction in the late 1950s [Carlin56]; when they were used primarily by microwave engineers. However, the availability of vector network analyzers has resulted in the regular use of these parameters for measuring devices at frequencies above 1 MHz, and most manufacturers now specify their devices in S-parameters.

The popularity of S-parameters for high-frequency analysis is related to problems in measuring devices:

- The open and short circuits required for Z- and Y-parameters cannot truthfully be implemented at any frequency, and implementation becomes problematic with increasing frequency.

- Attempting open- or short-circuit terminations on a device often allows the device to oscillate at some frequency above or below the desired measurement frequency.

- Voltage and current probes close to a device add capacitance or inductance, thus creating errors in measurement.

You can lessen these problems by placing the device in a precision coaxial line and then measuring reflected and transmitted waves. The resultant measurements are named S-, or scattering, parameters. S-parameters are as useful at optical frequencies as they are at radio frequencies (RF).

So far, we have used matrices for ease of automation and computation, but laboratory measurements create another need for matrices. Each of the matrix methods described so far has been based on ideal voltage and current sources. The use of ideal sources creates a measurement problem since real voltage or current sources have finite impedances that cause errors in the voltages and current measured. Futhermore, active devices often oscillate if their source and or load impedances are too high or too low. For these and other reasons, we want to measure devices with finite source and load resistances: 50 or 75 Ω terminations are commonly used. Scattering parameters conveniently require source and load terminations.

Scattering parameters consider each signal to be a wave. Some of the wave travels through the circuit, and some of it is reflected, or scattered, back to the source. A nonzero reflection of a wave means that maximum power is not being delivered from the source to the load: The difference between the maximum power and the delivered power is considered to be reflected back into the source.

The necessary condition for maximum power delivery is easy to understand when we consider that maximum power is delivered to a load only when the load impedance is the complex conjugate of the source impedance.

4.6.1 Reflections and Matched Terminations

We can prove in just a few steps that a conjugate match is required for maximum power transfer. This aside should give you an idea of how easy it is to use *Mathematica* in a symbolic mode. Given a source with impedance **Zs** connected to a load of impedance **ZL**, we can compute the current, compute the load voltage, and finally find the power dissipated in the load. The power dissipated is the real part of the product of the load voltage and the complex conjugate of the current through the load. As explained in Section A.7, the **ReIm** package included with *Mathematica* takes a prohibitively long time to solve the problems below. We will simplify our complex expressions with the functions presented in Section A.7. We compute the real part of the load power:

```
In:
  Zs = Rs + I Xs;
  ZL = RL + I XL;
  Current = Vs/(ZL+Zs);
  VL = Current ZL;

  realPower = getReal[VL getConj[Current]];
  Simplify[ realPower ]
```

```
Out:
                  2
           RL Vs
   ---------------------------
           2            2
   (RL  + Rs)  + (XL  + Xs)
```

By differentiating **realPower** with respect to **XL** and setting the result to zero, we can solve for an optimal value of **XL**:

```
In:
    Solve[ D[realPower, XL] == 0, XL]
```

```
Out:
    {{XL -> -Xs}}
```

Thus, the transferred power is maximized when **XL** = −**Xs**. If we then differentiate with respect to the load resistance and solve for the case when the derivative is zero, we can determine whether the function's stationary point is at a maximum or minimum:

```
In:
    Solve[ D[realPower /.
                  XL -> -Xs, RL] == 0, RL]

Out:
    {{RL -> Rs}}
```

If we inspect the formula for real power, a minimum of zero power will occur when **RL** is zero. So, this last result must be a maximum: Maximum power is delivered when the source and load are conjugates — the load is **Rs - I Xs**. Using the functions and symbolic capability of *Mathematica* has helped us solve the problem with a natural flow.

4.6.2 Use of S-Parameters in Analysis

S-parameters are defined in terms of waves that transmit into and reflect from each port of a network. A port is formed by a network node and ground. Each S-parameter is measured with the all ports terminated. These terminations form the port normalization vector for the *S* matrix. In *Nodal*, the terminations — usually resistors — are added automatically to the network in the S-parameter description but are not included in the other matrix descriptions. Thus you must always specify an impedance when transforming to or from S-parameters: 50 Ω is the default. As we said, a *Y*-matrix must have termination resistors added for its voltage and current results to agree with those of S-parameter analysis. These termination resistors are shown in Figure 4.7 as being external to the network described by the matrix. (You must also add resistors to SPICE analyses to derive the S-parameters directly from the voltages and currents normally given in SPICE results.)

The equation $b = S\,a$ describes how the *S* matrix is constructed. As shown in Figure 4.7, *a* is the incident wave vector and *b* is the reflected wave vector. The *S*-matrix, *S*, relates the incident and reflected waves.

As an analogy to optics, consider the *S*-matrix for a window with a lamp in a room as a source, a_1. Multiplying out the matrix, we find that $b_1 = S_{11}\,a_1$ and $b_2 = S_{21}\,a_2$. Most of the light from the lamp is seen outside, at port 2, and this gives b_2. So S_{21} becomes the light transfer from inside the room to outside. Some of the light is reflected by the window back into the room since someone sitting in the room can see an image of the lamp. The reflected light is represented by b_1, because port 1 is considered to be the inside of the room. Thus S_{11} represents the window reflection and S_{21} gives the window transmission.

All current S-parameter test equipment is based on two ports. You can measure multiport networks by terminating unmeasured ports and building the complete matrix by successive measurements.

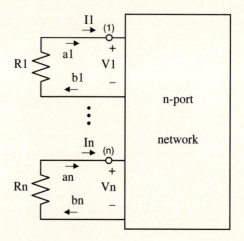

Figure 4.7 S-parameter problem definition.

The waves defined by a_i and b_i are actually combinations of voltages and currents: Each wave has its voltage-to-current ratio defined by the characteristic impedance of the line, and the voltages and currents on the line can be completely defined in terms of the incident and reflected waves on the line. The amplitude and phase of the waves are variables we can use when relating the waves to the voltages and currents. The wave amplitudes and phases are given by the S-parameters.

Consider node 1 in Figure 4.7. Taking node 1 as port 1, we can express V_1 and I_1 as the sum of incident and reflected wave voltage and currents.

```
In:
    port1Eqns =   {V1 == V1i + V1r,
                   I1 == I1i - I1r};
```

Because the reflected current is moving in the direction opposite to that of I_1 in Figure 4.7, it is given a negative sign. Now, we need to derive an equation of the form

$$V_{1r} = S_{11} V_{1i}$$

to match the form of the S-parameter matrix. We can solve for S_{11} in terms of V_1 and I_1 once we pick characteristic impedances for our ports. Traveling waves are difficult to imagine when only a single node is shown. It helps to imagine a transmission line of characteristic impedance R1 extending from R1 to node 1. The characteristic impedance for port 1 is R1; for simplicity, let us assume that all the ports have a characteristic impedance of **Ro**. Thus, **V1i/I1i = Ro** and **V1r/I1r = Ro**.

By substituting for the currents, we can arrive at the voltage based S-parameters by solving for **V1r** and **V1i**:

```
In:
   soln = Solve[ (port1Eqns /.
                  {I1i->V1i/Ro,I1r->V1r/Ro}),
                 {V1i,V1r}]

Out:
              I1 Ro   V1          -(I1 Ro)   V1
   {{V1i -> ----- + --, V1r -> -------- + --}}
              2      2            2        2
```

The incident and reflected voltages involve the measured voltages and currents as well as the characteristic impedance. We calculate **s11** from the ratio of **V1r** to **V1i**:

```
In:
   s11 = Simplify[V1r/V1i /. soln]

Out:
   -(I1 Ro) + V1
   {-------------}
    I1 Ro + V1
```

Although this answer is correct, it is more common to recognize V_1/I_1 as the input impedance of port 1, and so we substitute $V_1 = I_1 Z_{in,1}$ into the equation for **s11** to eliminate V_1:

```
In:
   S11 == Simplify[s11 /. V1->I1 Zin1]

Out:
          -Ro + Zin1
   S11 == {----------}
          Ro + Zin1
```

We note several special cases: What is the value of S_{11} when the input impedance is zero, equal to the source impedance, infinite, or minus twice the source impedance?

We can calculate the answer for each of these cases using an unnamed user-defined function that produces a table with two columns (**Zin1** and the corresponding **s11**):

```
In:
    tmp = {#,Limit[%[[2,1]], Zin1->#]}& /@
                            {0,Ro,Infinity,-2 Ro};
    TableForm[tmp,
            TableHeading -> {None, {{Zin1,S11}}}]
Out:
    Zin1        S11
    0           -1
    Ro          0
    Infinity    1
    -2 Ro       3
```

So, for short and open circuits, S_{11} is –1 or +1, respectively. No reflection occurs when the input is equal to the characteristic impedance, and only negative input resistances can produce reflections of magnitude greater than unity. All impedances with a positive real part are contained in a circle of unity radius in the reflection plane. If lines are drawn in the reflection plane for impedances of constant resistance or reactance a Smith chart is formed.

The **a** and **b** terms are proportional to the incident and reflected voltages, but they are usually presented in the normalized fashion shown here:

```
In:
    a1 == Simplify[V1i/Sqrt[Ro] /. soln]
Out:
                I1 Ro + V1
    a1 == {-----------}
                2 Sqrt[Ro]
```

```
In:
    b1 == Simplify[V1r/Sqrt[Ro] /. soln]
Out:
                -(I1 Ro) + V1
    b1 == {-------------}
                2 Sqrt[Ro]
```

Both **a1** and **b1** have units of **Sqrt[watts]**. So $|a|^2$ and $|b|^2$ have units of watts, which means that **a** and **b** represent power flow into and out of a network.

These equations show how, after calculating the port currents and voltages, you can derive the S-parameters for the network. Measuring port currents and voltages is how S-parameters can be derived for a known network. In practice, the incident and reflected waves are measured and then converted to port voltages and currents, thus allowing you to convert easily from other parameters, such as Z-parameters.

4.6.3 Conversion of Z-Parameters to S-Parameters

The formula for S_{11} has the same form as that for converting Z-parameters to S-parameters. Given that a Z-matrix is defined by

$$V = Z \cdot I,$$

where V and I are voltage and current vectors and Z is the impedance matrix of the network, we can expand $V = Z \cdot I$ to give

$$V_i + V_r = Z\,(I_i - I_r).$$

(Remembering that, for each port, $V_1 = V_{1i} + V_{1r}$ and $I_1 = I_{1i} - I_{1r}$.)

Because we also know that the incident and reflected wave voltages and currents are related by the characteristic impedances of the lines, then

$$V_i = R_o\,I_i$$

where R_o is now a diagonal matrix. Thus

$$R_o\,I_i + R_o\,I_r = Z\,(I_i - I_r).$$

Because this equation involves vectors and matrices, we cannot use **Solve**. However, a trivial rearrangement gives

$$I_i = (Z - R_o)^{-1} \cdot (Z + R_o) \cdot I_r.$$

So, the S-parameters can be derived from the Z-parameters and the port normalization impedances as

$$S = I_i/I_r = (Z - R_o)^{-1} \cdot (Z + R_o).$$

Rearranging the preceding equation gives

$$(Z - R_o)\,S = (Z + R_o)$$

and

$$Z\,(S - 1) = R_o + R_o\,S$$

which reduce to a formula for transforming the Z-matrix into an S-matrix:

$$Z = R_o \cdot (1 + S) \cdot (-1 + S)^{-1}$$

(The 1's and R_o's represent diagonal matrices.)

4.6.4 Properties of S-Parameters

S-parameters have several useful properties about which you should know and that you can demonstrate with *Mathematica*. First, a lossless line of characteristic impedance equal to the port impedance has an *S*-matrix of

$$\begin{bmatrix} S_{11} & S_{12} \\ S_{21} & S_{22} \end{bmatrix} = \begin{bmatrix} 0 & E^{-\Theta} \\ E^{-\Theta} & 0 \end{bmatrix},$$

where Θ is the line length in radians at the frequency being measured.

We do not change the magnitude of the S-parameters by placing a lossless line of characteristic impedance equal to the port impedance on any port because these lines have the same input impedance as a termination. Only the phase is changed. Because only the phase changes, this allows you to easily change the reference position via S-parameters: It is easy to measure at one end of a cable and to infer what has happened at the other end of the cable.

Although this remote measurement technique leads into error-correction and calibration issues with lossy and imperfect cables, you can site the test equipment separate from the devices under test. The remote measurement technique is very important when you want to measure, say, an antenna on a roof without hauling 50 kg of test equipment onto the roof, or device characteristics from –55°C to 85°C without baking your test equipment, or when you refer your results to a device only 15 mils wide using test equipment with ports several inches apart.

Often a network is measured in a 50 Ω system and used in a 75 Ω system. You can easily transform the measured S-parameters between impedance regimes as follows [Ha81]:

$$S_{\text{New}} = A^{-1} \bullet (S - \Gamma) \bullet (1 - \Gamma \bullet S)^{-1} \bullet A$$

where

$$A = R_{n\text{New}}^{-1} \bullet R_n \bullet (1 - \Gamma),$$
$$R_n = (R_o^{-1})^{1/2}/2, \text{ and}$$
$$\Gamma = (R_{o\text{New}} - R_o) \bullet (R_{o\text{New}} + R_o)^{-1}.$$

Note that A, R_n, Γ, and R_o are all diagonal matrices. When the terminations are complex impedances, the equations are more complicated.

You can convert the preceding equations to *Mathematica* and solve for any real impedance normalization. For a 2×2 transformation from 50 to 75 Ω, you have

```
In:
  (Ro=DiagonalMatrix[{50,50}]) //MatrixForm
```

Out:

```
50   0
0    50
```

```
In:
  RoNew = DiagonalMatrix[{75,75}];
  Rn = Sqrt[Inverse[Ro]]/2 //N;
  RnNew = Sqrt[Inverse[RoNew]]/2 //N;
  gamma = (RoNew - Ro).Inverse[(RoNew + Ro)];
  MatrixForm[gamma]
```

Out:

$$
\begin{matrix}
\dfrac{1}{5} & 0 \\[2mm]
0 & \dfrac{1}{5}
\end{matrix}
$$

```
In:
  A = Inverse[RnNew].Rn.(IdentityMatrix[2]-
                                  gamma);

  MatrixForm[A]
```

Out:

```
0.979796        0.
    0.        0.979796
```

Working through these equations results in a formula for the 50 to 75 Ω transformation. The original set of parameters is in a 50 Ω system, whereas the resulting set is in a 75 Ω system:

```
In:
  s = {{S11,S12},{S21,S22}};
  sNew = Simplify[Chop[Inverse[A] . (s-gamma) .
  Inverse[IdentityMatrix[2]-gamma.s] . A]];

  S11n75==N[ Together[Expand[sNew[[1,1]]]] ]
```

Out:

```
             -5. + 25. S11 + 5. S12 S21 + S22 - 5. S11 S22
S11n50==------------------------------------------------------
  -
             25. - 5. S11 - 1. S12 S21 - 5. S22 + S11 S22
```

```
In:
  S12n75==N[ Chop[Together[
               Expand[sNew[[1,2] ]]]] ]
```

```
Out:
                               24. S12
  S12n75 == -----------------------------------------------
            25. - 5. S11 - 1. S12 S21 - 5. S22 + S11 S22
```

```
In:
  S21n75 ==
          N[Chop[Together[Expand[sNew[[2,1]]]]] ]
```

```
Out:
                               24. S21
  S21n75 == -----------------------------------------------
            25. - 5. S11 - 1. S12 S21 - 5. S22 + S11 S22
```

```
In:
  S22n75 ==
          N[Together[Expand[sNew[[2,2]]]] ]
```

```
Out:
            -5. + 1. S11 + 5. S12 S21 + 25. S22 - 5. S11 S22
  S22n75==-----------------------------------------------
  --
            25. - 5. S11 - 1. S12 S21 - 5. S22 + S11 S22
```

Another important property of S-parameters is due to energy conservation. Because S-parameters are based on power flow, the properties of the matrices show conservation of energy principles. For example, if the network is lossless, then the power entering the network must be equal to the power leaving the network, or $|a|^2 = |b|^2$. Remember that a and b are vectors. Equating the squared magnitudes of a and b leads to the result that

$$S \bullet S^{t*} = I,$$

which means that the S-matrix in this case is unitary. I is the identity matrix in the above case. Also, the "^{t*}" operator creates the conjugate transpose of a matrix. When S is unitary the reflected and transmitted powers are linked as

$$|S_{11}|^2 + |S_{12}|^2 = 1$$

for a two-port circuit: The sum of the power reflected from port 1 and transmitted to port 2 equals the incident power on port 1. More practically, for a low-loss network, the loss determines the reflections, and a loss close to zero ensures small reflections and a good match. If the network is also reciprocal, $S_{12} = S_{21}$ and $|S_{11}| = |S_{22}|$.

4.7 Summary

In this chapter, you have seen how

- to analyze linear circuits by matrix techniques
- to build a circuit analysis program using admittance matrix techniques
- to transform between matrix forms (such as *ABCD* and *S*)
- to write functions that recognize data types
- to use *Mathematica* to manipulate matrices.

4.8 Exercises

4.1 Define a transformation from *Z*- to *G*-matrices. *G*-matrices use V_1 and I_2 as inputs and V_2 and I_1 as outputs. They are useful when parallel inputs and series outputs are used, *e.g.*, baluns.

4.2 Define a transformation from *Z*- to *H*-matrices. *H*-matrices use V_2 and I_1 as inputs and V_1 and I_2 as outputs. They are useful when series inputs and parallel outputs are used. *H*-matrices were popular in early transistor manufacture because h_{21} corresponds to the *beta*, or current gain, of the device.

4.3 Define a transformation from *S*- to *T*-matrices. *T*-matrices use a_1, b_1 as inputs and a_2, b_2 as outputs. They form cascade matrices for waves.

4.4 Solve the equations for a bridged-T network. Translate them to S-parameters. What is required for a bridged-T to provide a constant resistance?

4.5 Derive the *ABCD*-matrix for an *n*:1 voltage transformer.

4.6 Certain matrix representations work best for certain types of circuits and devices. What matrix is best for a T-pad? What matrix is best for a π-pad? What matrix best represents a field effect transistor (FET)?

4.7 Derive the *Y*-matrix of a single shunt element. The direct connection of two nodes creates problems in the *Y*-matrix implementation. How might you extend the *Y*-matrix to allow direct connections between two nodes and voltage sources to be used? (*Hint*: see [Vlach83].)

4.8 Show that the *S* matrix of a lossless network is unitary. That is,

```
Adjoint[S] == Inverse[S]
```

where

```
Adjoint[x_] := Transpose[Conjugate[x]]
```

4.9 Prove the equation for the change in port-normalization impedances of an S-parameter network.

4.10 Describe how you could write a translator function that would take a list of elements, nodal connections, and element values and return a Y-matrix or an *ABCD*-matrix. This function would be the core of an analysis program. A user could take a netlist, possibly from a schematic capture program, and convert it to a matrix that could be mapped over frequency.

4.11 Many circuit analysis problems are solved directly in the time domain. Modify the matrix loading and solving procedures of Section 4.5 to integrate a solution through time (*see* [Vlach83]).

4.12 Describe the problems that will arise in the matrix transformations of this chapter if the circuit is nonlinear, *i.e.*, the matrix values depend on voltage or current.

4.9 References

[Carlin56] Carlin, H.J., "The Scattering Matrix Approach in Network Theory," *IRE Trans. Circuit Theory,* pp. 88-96, June 1956.

[Ghausi65] Ghausi, M.S., *Principles and Design of Linear Active Circuits*, McGraw-Hill, New York, 1965.

[Ha81] Ha, T.T., *Solid-State Microwave Amplifier Design*, John Wiley & Sons, New York, 1981.

[Kuo75] Kuo, B.C., *Automatic Control Systems*, Prentice-Hall, Englewood Cliffs, New Jersey, 1975.

[Pohl67] Pohl, E., *Nachrichtentechnik*, Vogel-Verlag, 1967.

[Vlach83] Vlach, J., Singhal, K., *Computer Methods for Circuit Analysis and Design*, Van Nostrand Reinhold, New York, 1983.

[Withoff91] email from David Withoff.

[Wolfram91] Wolfram, S., *Mathematica – A System for Doing Mathematics by Computer*, Addison-Wesley, Redwood City, California, 1991.

CHAPTER 5

Component Design and Sensitivity Analysis

Although many components are available off the shelf, in some cases it is best that you design a custom component. For example, you might need an inductor of a particular value that must fit inside a specified space envelope or you might need a microstrip line of a specific impedance. Once you have designed the component, you will also need to know how the manufacturing tolerances affect the electrical performance of the component, as well as the production cost.

In this chapter we show you how *Mathematica* can help with the design task and the checking of component tolerances.

5.1 Component Value Functions and Utilities

Your first step in the solution of an electrical design problem is either to create your own function that returns the component's electrical property (capacitance, inductance, and so on) as a function of some physical parameter(s) (length, diameter, or gap) or to use some of *Nodal*'s utilities to help calculate the value of a component from a physical description. You can then use the output from your function or the utility to specify the component's value in the netlist. Because *Mathematica* can work symbolically, a parameter of a component can be specified in terms of another symbolic variable.

For example, you can calculate the value of a capacitor made from two parallel plates with physical dimensions 2.5 m long × 1 mm wide × 2 mm apart (in a dielectric constant of 1.00) by defining your own function **plateC**. Your function will evaluate the standard formula for the capacitance of two parallel plates of area **a** separated by a gap distance **d** in a medium of relative permittivity **er**, $C = \varepsilon_o$ **a er** / **d**, where ε_o is the permittivity of space (8.854×10^{-12} F/m). Alternatively, you can perform the same calculation by using *Nodal*'s **Capacitance** utility:

```
In:
    plateC[a_,d_,er_]:= 8.854 10^-12 a er /d

    plateC[2.5 10^-3, 2 10^-3, 1.0]
```

```
Out:
                -11
   1.10675 10
```

```
In:
   Capacitance[2.5, ParallelPlate[0.001,0.002]]
```

N

```
Out:
                -11
   1.10675 10
```

Functions that have physical dimensions as their arguments allow you to keep the electrical description of a component entirely general — you do not need to hard-code numerical values (for example, 11 pF) into your netlist or analysis. You can use functions to keep the value of a component as a function of some other variable: A variable capacitor can have the overlap of its plates defined in terms of a fluid level in a storage tank, for example. If the overlap of the plates is 18.34 mm multiplied by a measure of the level called **level**, and there is one vane 1 mm wide that is made from two plates 100 µm apart in a dielectric of permeability 1.0, then you can calculate the capacitance, in farads, as a function of **level**:

```
In:
   overlap=18.34 mm level;
   Capacitance[overlap,
                   ParallelPlate[1 mm, 0.1 mm, 1.0]]
```

N

```
Out:
              -12
   1.62382 10     level
```

Of course, you can use more complicated functions for the overlap. Using utilities enables you to keep problems object-oriented.

5.2 RLC Filter Design

Filters are probably the most common circuits that contain only resistors, inductors, and capacitors. Indeed, you can make quite complex filters without any active devices. The ability of filters to select or reject frequencies is fundamental to many devices — from TVs and radios to equipment that measures the most complex scientific experiments. You can use *Nodal* and *Mathematica* to help you design filters — and to visualize their behavior as a function of frequency.

In our next (rather long) example, we analyze a tuned circuit, determine the output–input voltage ratio as a function of frequency, and then proceed to design the required inductor — including considering how tolerances in the manufacture of the filter will affect the performance of the final product. First, let us define the circuit and use *Nodal* to calculate how the output varies with input frequency:

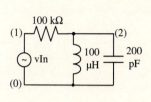

In:
```
tune=NodalNetwork[Resistor[{1,2},100 kOhm],
                  Inductor[{2,0},100 uH],
                  Capacitor[{2,0},200 pF],
                  VoltageSource[{1,0},vIn]];
NodalAnalyze[tune,Result->V2/V1]//Simplify
```

Out:

$$
\frac{-25000\ \text{Pi}\ f}{12500000000000\ I - 25000\ \text{Pi}\ f + -I\ \text{Pi}^2\ f^2}
$$

You can then define a function that calculates the performance of the tuned circuit, in terms of the magnitude and phase angle of the output signal (compared with the input). First, you need to extract the preceding equation from the enclosing list structure, using the **Normal** function, and assign a name, **vratio**:

In:
```
vratio=Normal[%][[1,2]];
Plot[Abs[vratio /. f->10^x],
     {x,1,8},
     PlotRange->All,
     AxesLabel->{"Log(f)","vratio"}]
```

Out:

vratio

[plot showing vratio vs Log(f) with a sharp peak near Log(f)=6]

This plot shows that the tuned circuit acts as a good attenuator except around 1 MHz, when its attenuation decreases dramatically. Although you have used

Nodal to compute the electronic circuit characteristics, *Mathematica* has produced this plot. *Nodal* has built-in plotting functions of its own that are designed to support engineering requirements rather than the more general plotting functions of *Mathematica*. (The subject of plotting and graphics is described in more detail in the Appendix.)

The Bode plot is a convenient, and often used, representation of gain and phase in the same diagram. To show how the output voltage of **tune** varies, we use *Nodal*'s graphics function **BodePlot** to produce a combined dB and phase (in degrees) plot as the frequency applied to **tune** ranges from 100 kHz to 10 MHz:

In:
```
BodePlot[NodalAnalyze[tune,
            Result->V2/V1,
            Frequency->(10^Range[-1,1,0.2] MHz),
            Step->10^Range[-1,1,0.2]]]
```

Out:

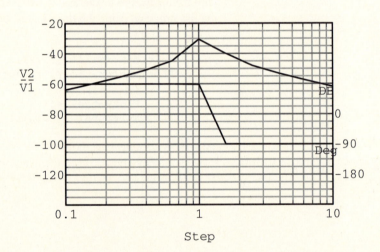

If you wish to design the 100 μH inductor used in **tune**, then you need to decide how it will be made — you could choose to make it from a parallel wire, ribbon, single- or multiple-layer solenoid, spiral, or toroid. *Nodal* is able to compute the inductance of all these structures, plus via holes (in printed circuit boards) and straight wire.

Choosing (arbitrarily) a 3 cm long, 1 cm diameter former for a single-layer solenoid, the simplest way to determine how many turns are required is to plot out the inductance as a function of the number of turns. By restricting the plot range, you confine the graph to your area of interest:

In:
```
Plot[Inductance[3 cm,
       SingleLayerSolenoid[1 cm,nTurns]]/uH,
      {nTurns,100,200},
      PlotRange->{95, 105},
      AxesLabel->{"turns","inductance"}]
```

Out:

Using this graphical approach, you can see that approximately 187 turns will give a 100 µH inductor — the graph also tells you that the inductance changes quickly as a function of the number of turns: The gradient of the line is very steep. Note that the inductance values are given in henries and that we then scaled them to microhenries by division by **uH**. *Nodal* always uses fundamental SI units.

A more elegant approach to the problem of calculating the number of turns is to use *Mathematica*'s root-finding function to search for zero points in a function. Root finding gives a more precise answer, is not subject to the problems of graphical interpretation, and can be a convenient way of determining the argument value that will cause a function to return a particular value:

In:
```
FindRoot[100 uH - Inductance[3 cm,
                  SingleLayerSolenoid[1 cm,nt]],
          {nt,200}]
```

Out:
```
{nt -> 187.63}
```

5.3 What-If Sensitivity Analysis

When you design a component, it is important — especially from a production engineering point of view — to understand how small deviations from the perfect specification will affect the electrical value of the component and also how real components vary from the idealized models that you may have used in your design calculations. In this section, we look at the simple what-if technique for assessing the effect of tolerances in the dimensions of an inductor on its inductance and how the resistance of the inductor's windings affects the performance of a simple RLC filter; we discuss a more analytical approach in Section 5.4.

You can assess the variation in inductance due to small changes in the diameter of the solenoid by using a very simple technique: Instead of using a single value for the diameter, insert a list of values that represent, say, the limits due to manufacturing tolerances. If the manufacturing process responsible for the diameter of the solenoid has a tolerance of ±0.1 mm, you can calculate the resulting inductance spread:

In:
```
Inductance[3 cm,
    SingleLayerSolenoid[({0.99,1.01} cm),187]]/uH
```

Out:
```
{97.9054, 101.636}
```

There are a few points to note here. First, you gave *Nodal* a list, `{0.99,1.01}`, instead of a single number. The two numbers in the list, `0.99` and `1.01`, are multiplied by `cm`, one of *Nodal*'s scaling factors. The function then calculates the solenoid's inductance for each value in the list — and so gives two answers, also in the form of a list. Dividing the list by `uH` gives the answer in microhenries rather than henries. By inspection, each of the inductances computed is ~1.8% in error from our desired value; the 1% error in diameter is nearly doubled when translated into inductance error.

How does the error in the inductance affect the frequency response of the filter? By generating two plots and then combining them, you can show the effect of the 1% change in solenoid diameter. If you incorporate a parameter `err` in the diameter of the solenoid and analyze the circuit symbolically, you will obtain a function that has `err` as one of its variables. When you alter the value of `err`, *Mathematica* only has to reevaluate the function — it does not have to reanalyze the circuit.

```
In:
  tune1=NodalNetwork[
          Resistor[{1,2},100 kOhm],
          Inductor[{2,0},
            Inductance[3 cm,
              SingleLayerSolenoid[1 err cm,187]]],
          Capacitor[{2,0},200 pF],
          VoltageSource[{1,0},vIN]];
  tuneFunc=Normal[NodalAnalyze[tune1,
              Result->V2/V1]][[1,2]];
```

You can then use **tuneFunc** as a function — with frequency, **f**, and manufacturing error, **err**, as parameters — that you can plot to see the effect of the manufacturing error:

```
In:
  Plot[Abs[tuneFunc /. {f->10^x,err->0.99}],
       {x,5.9,6.1},
       AxesLabel->{"Log(f)","response"}]

Out:
```

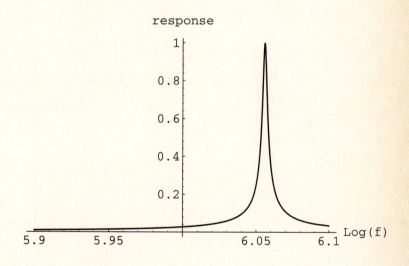

```
In:
  Plot[Abs[tuneFunc /. {f->10^x,err->1.01}],
       {x,5.9,6.1},
       AxesLabel->{"Log(f)","response"}]
```

Out:

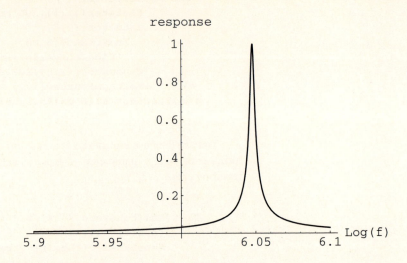

You can then display and compare both plots on the same graph using the **Show** command:

In:
    Show[%,%%]

Out:

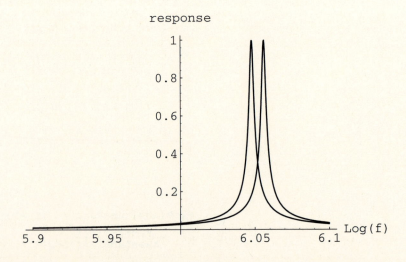

You can determine how changes in the length (say, ±10%) and the number of turns (±1) affect the inductance in an identical way:

```
In:
    Inductance[({2.9,3.1} cm),
        SingleLayerSolenoid[(1 cm), 187]]/uH
Out:
  {102.741, 96.9528}

In:
    Inductance[3 cm,
        SingleLayerSolenoid[(1 cm), {186,188}]]/uH
Out:
  {98.6989, 100.833}
```

The frequency selectivity of a tuned circuit decreases as the circuit's resistance increases, so it is useful to be able to calculate the resistance of the solenoid's wire. When you make the solenoid, you also need to know what thickness of wire to use and how long a length to cut off the storage reel. You can collate these three calculations into a simple function. We have placed the body of the function into a **Module** with declared variables to constrain the scope of the variables **wireDiameter** and **wireLength** to our function:

```
In:
solenoidFacts[solenoidLength_,
                numTurns_,
                solenoidDiameter_]:=
Module[{wireDiameter, wireLength},
  wireDiameter=solenoidLength/numTurns;
  Print["diam (microns) =",
        wireDiameter/micron];
  wireLength=Pi solenoidDiameter numTurns;
  Print["total wire length =",
        wireLength,"meters"];
  Print["resistance (Ohms) =",
        Resistance[wireLength,
                    Wire[wireDiameter]]//N];]

solenoidFacts[3 cm, 187, 1cm]
Out:
  diam (microns) = 160.428
  total wire length = 5.87478 meters
  resistance (Ohms) = 4.99886
```

You can calculate the resistance for a particular conductor using *Nodal*'s **Resistance** utility. For example, a 5.87 m length of wire 160 μm in diameter can have its resistance calculated as follows. *Nodal* assumes that the wire is copper and is being used at DC. *Nodal* provides two resistivity constants: copper and gold.

```
In:
   Resistance[5.87,
                 Wire[160 micron]]
```
N

```
Out:
   5.02154
```

When a conductor is used at high frequencies, skin effect becomes important and will raise the value of the conductor's effective resistance. Skin effect can be calculated by *Nodal*:

```
In:
   SkinDepth[Rho[Copper],1 MHz]
```
N

```
Out:
       0.000206155
       -----------
            Pi
```

When you use **Resistance**, skin effect can also be included automatically by specifying the frequency at which the resistance is to be calculated:

```
In:
   Resistance[5.87,
                 Wire[160 micron],
                 Frequency->1 MHz]
```
N

```
Out:
   5.52508
```

For other materials, you can specify their resistivity in the conductor specification (**Wire** in this example):

```
In:
   Resistance[5.87,
              Wire[160 micron, Rho->95 10^-8],
              Frequency->1 MHz]
```
N

```
Out:
   293.861
```

You may wish to review and change the coil design to take advantage of a particular off-the-shelf wire gauge. The preceding example assumed a wire diameter of 160 μm; this diameter can be converted to mil so that you can make a comparison with AWG values:

```
In:
   160 micron / mil
```
N

```
Out:
   6.2992
```

The diameter value of 6.2992 mil is very close to the diameter of AWG34 wire. However, if only 8 mil diameter wire were available, you would want to rerun the coil calculations. How will the adoption of the standard wire gauge affect the length and resistance of the solenoid? The length of the single-layer solenoid in centimeters (the length of the solenoid, not the wire length) is simply the product of the number of turns and the diameter of the wire:

```
In:
  diamWire=8 mil;
  FindRoot[100 uH -
    Inductance[nTurns diamWire,
                SingleLayerSolenoid[1 cm,
                                    nTurns]],
          {nTurns,200}]
Out:
  {nTurns -> 226.626}

In:
  length=(nTurns diamWire /. %)/cm
Out:
  4.60505
```

You can now compute the length of wire required and the resistance, knowing that AWG32 wire (corresponding to 8 mil diameter) has a resistance of 167.3 Ω per 1000 feet:

```
In:
  wirelength=nTurns Pi 1 cm /. nTurns->226.6
          (*in meters*)
Out:
  2.266 Pi

In:
  resistance=N[wirelength 167.3 / (1000 feet)]
Out:
  3.90742
```

Alternatively, you can use *Nodal*'s **Wire** and **Resistance** utilities:

```
In:
  N[Resistance[2.266 Pi, Wire[8 mil]]]
Out:
  3.77572
```

Recalculating the resistance of the wire allows you to check the effect of the inductor's resistance on the frequency selectivity of the filter. (Note the softer peak.)

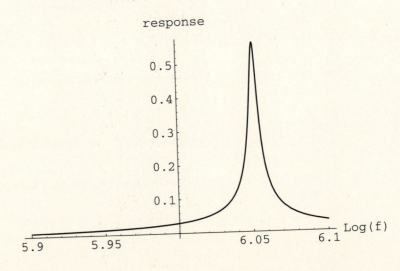

```
In:
    newFilter=NodalNetwork[VoltageSource[{1,0},vIn],
                Resistor[{1,2},100 kOhm],
                Inductor[{2,3},100 uH],
                Resistor[{3,0},3.9],
                Capacitor[{2,0},200 pF]];
    NodalAnalyze[newFilter,
                Result->V2/V1]
```

```
Out:
                                           I
                                          --
                                 -6      20
                        -2.5641 10    + ----
                                        f Pi
    --------------------------------------------------
                    -6     1282.1 I                 -10
        -4.5641 10     + --------- - 1.02564 10      I f Pi
                          f Pi
```

```
In:
    Plot[Abs[Normal[%][[1,2]] /. f->10^x],
        {x,5.9,6.1},
        AxesLabel->{"Log(f)","response"}]
```

Out:

5.4 Differential Sensitivity Analysis

The what-if technique is simple, but there is an alternative that retains a mathematical — rather than arithmetic — approach and that is amenable to symbolic manipulation: *differential sensitivity analysis*.

Given a quantity y that is a function of x (that is $y = f(x)$), then when x changes by a small amount, dx, the corresponding change, dy, in y is given by $dy = f'(x)\,dx$, where f' is the derivative of f. Because *Mathematica* is able to compute derivatives of functions, it can help you perform differential sensitivity analysis.

For example, the voltage output from a simple potential divider circuit is

```
In:
   vfrac=r2/(r1+r2)

Out:
       r2
    -------
    r1 + r2
```

You can obtain the partial derivatives with respect to **r1** and **r2** of **vfrac** using the differentiate function **D**. You can then evaluate the numeric values of the partial derivatives for the values of **r1** and **r2**:

```
In:
   D[vfrac,r1]

Out:
             r2
     -(-----------)
                 2
        (r1 + r2)


In:
   % /. {r2->1000, r1->1000}

Out:
         1
     -(----)
       4000


In:
   D[vfrac,r2]

Out:
             r2              1
     -(-----------)  +  -------
                 2        r1 + r2
        (r1 + r2)
```

```
In:
  % /. {r2->1000, r1->1000}
```

```
Out:
      1
     ----
     4000
```

You can then calculate the change, **vOutDelta**, in the potential divider's output voltage (375 µV) that results from an increase of 1.5 Ω in **r2**:

```
In:
  r2delta=1.5;
  vOutDelta=(D[vfrac,r2] /. {r1->1000,
                             r2->1000}) r2delta
```

```
Out:
  0.000375
```

If both **r1** and **r2** are subject to errors δ**r1** and δ**r2**, respectively, then the *change* in the output voltage δ**vOut** is given by

$$\delta\mathbf{vOut} = \frac{\partial}{\partial\mathbf{r1}}(\mathbf{vfrac})\,\delta\mathbf{r1} + \frac{\partial}{\partial\mathbf{r2}}(\mathbf{vfrac})\,\delta\mathbf{r2}$$

You can easily calculate **dvOut**:

```
In:
  r1delta=1.0;
  r2delta=1.5;
  dvOut=(D[vfrac,r2] r2delta +
            D[vfrac,r1] r1delta) /.
                     {r1->1000, r2->1000}
```

```
Out:
  0.000125
```

If you want the actual output voltage, rather than the change in it, then you just add the change to the error-free voltage:

```
In:
  r1delta=1.0;
  r2delta=1.5;
  (vfrac +
     D[vfrac,r2] r2delta +
         D[vfrac,r1] r1delta) /.
                     {r1->1000, r2->1000}
```

```
Out:
  0.500125
```

You can use differential sensitivity analysis to help improve a design by finding the conditions that minimize the errors created by tolerance and drift (which may be due to temperature changes or aging) in component values. For example, by returning to the symbolic form of **dvOut**, you can find the necessary conditions under which **dvOut** will be zero:

In:
```
Clear[r1, r2, r1delta, r2delta, length];
dvOut=(D[vfrac,r2] r2delta +
            D[vfrac,r1] r1delta)
```

Out:
```
    r1delta r2                 r2            1
 -(----------) + (-(----------) + -------) r2delta
        2                   2      r1 + r2
    (r1 + r2)          (r1 + r2)
```

In:
```
Solve[dvOut==0,{r1delta,r2delta}]
```

Out:
```
                r1 r2delta
    {{r1delta -> ----------}}
                    r2
```

So, if **r1** is 100 Ω and **r2** is 1 kΩ, any change in **r1** has to be one-tenth that in **r2** for zero error in the output of the potential divider.

You can often obtain a qualitative feel for how parameters will affect performance by inspecting a symbolic differential. How is the error in a single-layer solenoid's inductance affected by changes in its length? By using *Nodal*'s **Inductance** function, you can obtain a formula for the solenoid's inductance, which you can then differentiate with respect to its length:

In:
```
h1=Inductance[length,
    SingleLayerSolenoid[diam,turns]]
```

Out:
```
            -6    2     2
 9.8425 10   diam  turns
 -----------------------
   4.5 diam + 10. length
```

In:
```
D[h1,length]
```

```
Out:
                    2        2
     -0.000098425 diam   turns
     -------------------------
                            2
     (4.5 diam + 10. length)
```

By inspection, you can see that errors in length will have a small effect on the inductance providing the length is small compared to the diameter or the length is large.

5.5 Cost Minimization

Product cost is one aspect of any design with which engineers must be familiar. If you know which factors drive the cost of a design, then you can make alterations to a product early in its design phase to optimize the performance–cost ratio. The result is not just a good product from the engineer's point of view, but also a better product from the client's viewpoint because it has cost less, too.

Mathematica can help you in product cost evaluation. In the following example, a 100 µH inductor has to be made. For a given inductance, a range of coil diameters and lengths can be used, and the particular combination of diameter and length used to implement the design will affect the cost. We tackle the problem by defining a function that takes three arguments (the required inductance, coil length, and coil diameter) and returns the cost.

First, you find the number of turns (the free variable): A convenient way to proceed is to use **FindRoot** to calculate numerically the number of turns required by determining when the value returned by *Nodal*'s **Inductance** function equals the required coil inductance, **cInductance**. The result of the **FindRoot** function is a rule that specifies the value assigned to the number of turns variable, **nT**. We have assigned the rule the name **nTrule**.

The cost of the complete coil assembly consists of three components: the costs of the wire, the bobbin, and the winding process. The bobbin is bought from a contractor, from whose catalog you find that each one costs 3¢ plus 1.5¢ per cm of length (**cLength**) plus another factor that is roughly 3¢ per cm radius (**cDiam/** 2) squared. You may have found analytical expressions for these costs by using *Mathematica* to fit a function (say, constant plus terms in length and radius) to costs from the contractor's catalog.

The wire cost is somewhat more simple — the wire catalog implies that it is really only the total weight of wire that affects the price ($30/kg) — which you can calculate by working out the diameter of the wire, **wDiam**, equal to the length of the coil divided by the number of turns, the total length of wire required, **wLength**, and knowing the density of the wire material.

The winding cost is worked out to be 1¢ per 100 turns for 1 mm diameter wire; thinner wire becomes more expensive because thin wire is more difficult to handle:

```
In:
  cost[cInductance_,cLength_,cDiam_]:=
  Module[{nTrule,nTurns,wDiam,costBobbin,
          wLength,wArea,densityCopper,
          wWeight,costWire,costWinding,
          totalCost},
    nTrule=FindRoot[cInductance ==
            Inductance[cLength,
             SingleLayerSolenoid[cDiam,nT]],
            {nT,500}];
    nTurns=nT /. nTrule;
    wDiam=cLength/nTurns;
    costBobbin=3 + 1.5 cLength/cm +
                         3 (cDiam/cm)^2;
    wLength=nTurns Pi cDiam;
    wArea= Pi (wDiam/2)^2;
    densityCopper=8900;
    wWeight=wLength wArea densityCopper;
    costWire=3000 N[wWeight];
    costWinding=(nTurns/100) (0.001/wDiam);
    totalCost=costBobbin+costWire+costWinding
  ]
```

You can then calculate the cost of a typical coil (in cents), just as you would use any *Mathematica* function:

```
In:
  cost[100 uH, 1 cm, 1 cm]
Out:
    22.7748
```

To obtain a better view of how the cost varies with length and diameter, you can plot out the cost as a function of these two variables:

```
In:
  Plot3D[cost[100 uH, length, diam],
         {length, 1 cm, 4 cm},
         {diam, 1 cm, 3cm},
         PlotPoints->5,
         AxesLabel->{"length cm",
                 "    diam cm","cost  "}]
```

Out:

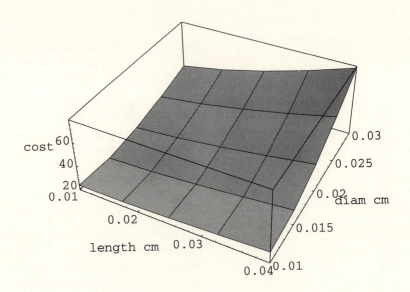

The graphical representation of the cost clearly shows that it varies relatively little with length but that it is a much stronger function of radius.

5.6 Summary

In this chapter, you have seen how

- to use user-defined functions or *Nodal* utilities to compute component electrical values from the physical specification of the component
- to determine component physical parameters given a desired electrical value by graphical and root-finding techniques
- to calculate the effects of manufacturing tolerances on electrical values using simple what-if and analytical techniques
- to estimate and to minimize costs of manufacture.

5.7 Exercises

5.1 For an inductance of 100 μH, compare the geometrical dimensions required for single-layer and multilayer solenoids, spiral, and toroidal inductors.

5.2 At what frequencies does the inductance of via holes become important? How do you decide what is "important"?

5.3 How close would two 1 m square parallel plates have to be to have a capacitance of 1 F? What permittivity did you choose? What would be the maximum voltage that you could apply across the capacitor?

5.4 High and medium values of inductances are obtained by winding coils on materials with permeabilities greater than that of air. What are typical values for ferrite permeabilities, and how does their use alter the dimensions of, say, a 5 mH inductor?

5.5 Write a program that computes the wire gauge, number of turns, diameter, and length of an inductor for a specified value of inductance and give some guidance about the size. The program should use only standardized wire gauges (SWG or AWG).

5.6 A high-voltage capacitor has to work at 1 kV. A value of 10 mF is required. Choose two types of dielectric material, and compute the cost of the capacitor as a function of the geometry.

5.7 For the capacitor in Exercise 5.6, determine how sensitive its value is to errors in the geometry that might occur during manufacture. Given that the cost of manufacturing equipment has to be amortized in the component cost, how would you estimate costs that might result from using tight tolerances?

5.8 Microstrip line components can be made by etching the appropriate pattern on a printed circuit board or hybrid substrate. Estimate the absolute tolerances (in millimeters) required for 5% variation in impedance.

CHAPTER 6

Time Series and Spectral Analysis

Historically, the analysis of time series has not been considered to be part of electronic engineering. The increasing use of digital signal processing, however, brings most engineers into contact with time series sooner or later. We devote the early part of this chapter to time-series techniques, and then we examine synthesis and decomposition of functions and basic frequency-domain filtering using Fourier methods. We cover digital filters — important tools for spectral analysis in the time domain — in Chapter 8.

6.1 Time Series

Mathematica is equally at home working on discrete numbers and on continuous mathematical functions. The core of *Mathematica* provides basic facilities for handling series of numbers (which are identical to *Mathematica*'s lists), and the standard package collection, which is supplied with *Mathematica*, expands the core facilities to include many graphing and statistical functions. In this section, we introduce you to handling time series. (Reading from and writing to external data files is described in the Appendix.)

6.1.1 Generation of Time Series

Time series can be generated in two ways: as the results of an experiment that you have performed or of executing a function, within *Mathematica*, that manufactures the series. The simplest way to create a series is to enter numbers in the format of a list. You can assign a name, such as **data1**, to a list using **=**:

```
In:
  data1={1,2,3,4,5,1,7,1,2}

Out:
  {1, 2, 3, 4, 5, 1, 7, 1, 2}
```

For longer series, series that you can create from a function, or multidimensional series, you can use **Table**. When you specify the series expression in **Table**, its form must mimic the structure in the required series: To create a table of number pairs, you enclose the output expression in **{}**. Applying the enumerate function, **N**, to the output expression forces *Mathematica* to format the output as real numbers. Here, we have specified that we want the numerical output formatted with three digits of precision:

In
```
data2=Table[N[{x,Sin[x]},3],{x,0,9}]
```

Out:
```
{{0, 0}, {1., 0.841}, {2., 0.909}, {3., 0.141},
{4., -0.757},{5., -0.959}, {6., -0.279}, {7., 0.657}, {8.,
0.989}, {9., 0.412}}
```

6.1.2 Statistical Analysis and Plotting

Once your data have been read in or generated, you can use *Mathematica*'s standard plotting functions to visualize the data. For a one- or two-dimensional list of numbers, **ListPlot** will generate a straightforward plot of the data:

In:
```
dataList={1,3,5,7,8,8.5,8.6};
ListPlot[dataList,
         PlotLabel->"regular x-spacing",
         AxesLabel->{"x","value"}]
```

Out:

In:
```
scatterList={{1,1},{8,2},{8.5,3},{9,10.2}};
ListPlot[scatterList,
         PlotLabel->"scatter graph",
         AxesLabel->{"x","value"},
         AxesOrigin->{0,0}]
```

Out:

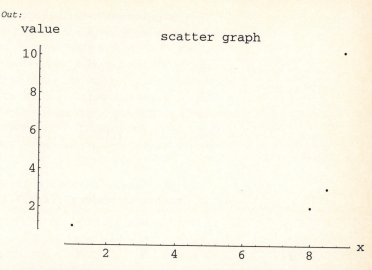

For more complicated plots, you can use the standard packages in the **Statistics** and **Graphics** families to analyze and plot data in a wide variety of ways. We strongly recommend that you take a few minutes to browse through the booklet *Guide to Standard Mathematica Packages*, which came with your copy of *Mathematica. Mathematica* may appear not to have such facilities as filled plots, labeled point plots, pie charts, bar graphs, error-bar plotting, polar plots, and three-dimensional graphs, but you will find them in the **Graphics** packages. Similarly, the **Statistics** packages contain many analysis functions for continuous and discrete distributions. Functions from these two packages work well together.

You can load a package into *Mathematica* using the **Get** function (with prefix form **<<**) before you can access the functions that it contains. Here are a few simple examples of elementary series analysis from the package **Statistics`DescriptiveStatistics`**:

In:
```
<<Statistics`DescriptiveStatistics`
Mean[data1]
```

Out:
```
26
--
9
```

In:
```
{Mode[data1], Median[data1]}
```
Out:
```
{1, 2}
```

The data-manipulation package contains functions that can manipulate data tables that have been read in. For example, they can select certain columns, drop nonnumeric data, and select items from the read-in list according to criteria that you specify. This package also contains functions to analyze the frequency of occurrence of values, to bin data, and to create lists of items that pass a specified test. Here is an example that analyzes **data1** for the number of items less than 1, between 1 and less than 3, between 3 and less than 8, and greater than or equal to 8; then analyzes the frequency of occurrence; and finally plots a bar chart from the frequency data (which are not evenly spaced):

In:
```
<<Statistics`DataManipulation`
RangeCounts[data1,{1,3,8}]
```

Out:
```
{0, 5, 4, 0}
```

In:
```
freq=Frequencies[data1]
```

Out:
```
{{3, 1}, {2, 2}, {1, 3}, {1, 4}, {1, 5}, {1, 7}}
```

In:
```
<<Graphics`Graphics`
BarChart[freq,
        AxesLabel->{"value",
                        "occurrence\ncount"}]
```

Out:

6.1.3 Generation of Time Series with Specific Noise Properties

As well as series of a deterministic nature, you can generate series of random values. These series are useful for testing algorithms and for adding noise to artificial data to give the latter realism. Once you have generated the series, you can export it from *Mathematica* to an external file, so that it can be used by other packages. For example, you might want to feed it through a digital-to-analog converter that drives a part of your experiment. In the next example, we create a series of 500 random numbers and analyze the frequency of values by plotting a bar chart.

Toolbox

Random-Number Generation

Random[] returns a random real number in the range 0 to 1.

Random[*type*, *max*] and **Random[*type*, {*min*, *max*}]** return random numbers between 0 and *max* and *min* and *max*, respectively, of the specified *type*. *type* can be one of **Real**, **Complex**, or **Integer**. An optional third argument to **Random** specifies the precision for the result.

Use **SeedRandom[]** to reseed the random-number generation; an optional argument can be used to specify the seed if you require a repeatable random sequence.

```
In:
  rand=Table[Random[],{i,1,500}];
  binCount=BinCounts[rand,{0,1,0.2}]

Out:
  {65, 43, 40, 49, 42, 55, 50, 56, 54, 46}

In:
  xcoords=Table[x,{x,0.1,0.9,0.2}]

Out:
  {0.05, 0.15, 0.25, 0.35, 0.45, 0.55, 0.65, 0.75, 0.85,
  0.95}
```

In:

```
BarChart[Transpose[{binCount,xcoords}],
    AxesLabel->{"mid-bin\nvalue",
                "occurrence\ncounts"}]
```

Out:

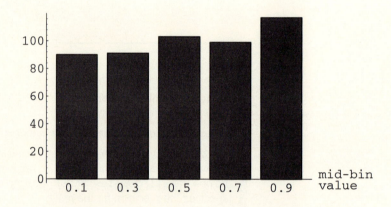

The packages **Statistics`ContinuousDistributions`** and **Statistics`DiscreteDistributions`** contain a selection of distributions that you can use with **Random** to produce colored random numbers.

For example, you can generate random numbers from a Poissonian distribution with a mean of 3 and visualize the distribution using a bar chart that displays the frequency of occurrence in equispaced bins:

In:

```
<<Statistics`ContinuousDistributions`
<<Statistics`DiscreteDistributions`

pRand=Table[Random[PoissonDistribution[2]],
            {i,1,500}];
bc=BinCounts[pRand,{0.5,5.5,1}];
xc=Table[x,{x,1,5,1}];
BarChart[Transpose[{bc,xc}]]
```

Out:

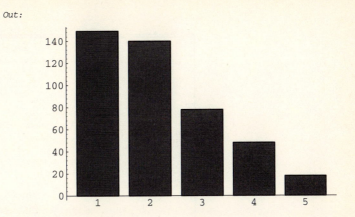

If your computer system supports the sound facilities in *Mathematica*, you can compare aurally **rand** and **pRand** by playing the lists:

In:
```
ListPlay[rand,
          SampleRate->500,PlayRange->{0,1}]
```

Out:

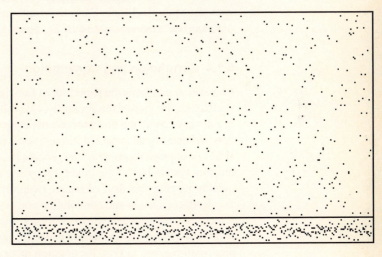

-Sound-

In:
```
ListPlay[pRand,
          SampleRate->500, PlayRange->{0,1}]
```

Out:

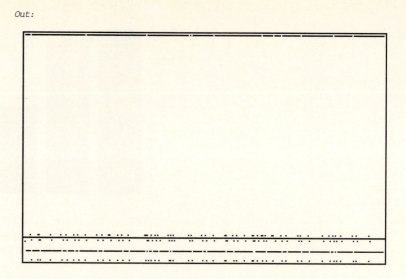

    -Sound-

Further functions for generating prescribed-distribution random numbers for use
with, for example, Monte Carlo techniques are given in the package
Statistics`InverseStatistical Functions`.

6.1.4 Correlation and Convolution

Correlation measures the similarity of two time series; the larger the correlation
coefficient, the greater the similarity. Convolution gives a running weighted sum
of one series, with the other series providing the weights for the summation process. When a signal is passed through some process, the point spread function
(PSF) or impulse response of the process is convolved with the signal: A slow-
reacting system has a long impulse response and so is incapable of reacting to
quickly varying signals [DeFatta88, Lynn82].

   Both correlation and convolution consist of summing a set of products of
selected elements in the two series. In conventional programming languages, you
would use two loops, one nested inside the other, as part of the algorithm. *Mathematica* allows you to keep the notation more mathematical: The result of either
correlation or convolution is a table of summations, with each entry in the table
corresponding to a particular group of selected elements.

   It is straightforward to carry out either process with *Mathematica* using
Table and **Sum**, but calculation of convolution and correlation in the time domain
is slow because of the large number of arithmetic operations involved. Neither
correlation nor convolution has an elegant algorithm. Here are two procedural

functions for calculating correlations and convolutions on a data array **d**, with
point spread function in the array **p**:

```
In:
  datax=Table[d[i],{i,5}]
```
```
Out:
  {d[1], d[2], d[3], d[4], d[5]}
```
```
In:
  psfx=Table[p[i],{i,3}]
```
```
Out:
  {p[1], p[2], p[3]}
```
```
In:
  Table[Sum[datax[[j]] psfx[[Length[psfx]+j-i]],
           {j,
             Max[1,i-Length[psfx]+1],
             Min[i,Length[datax]],1}],
         {i,1,Length[datax]+Length[psfx]-1,1}]
```
```
Out:
  {d[1] p[3],
   d[1] p[2] + d[2] p[3],
   d[1] p[1] + d[2] p[2] + d[3] p[3],
   d[2] p[1] + d[3] p[2] + d[4] p[3],
   d[3] p[1] + d[4] p[2] + d[5] p[3],
   d[4] p[1] + d[5] p[2],
   d[5] p[1]}
```

Convolution is performed in the same manner except that the array that is
convolved onto the data is reversed prior to being used:

```
In:
  Table[Sum[datax[[j]] psfx[[i-j+1]],
  {j,Max[1,i-Length[psfx]+1],Min[i,Length[datax]],1}],
  {i,1,Length[datax]+Length[psfx]-1,1}]
```
```
Out:
  {d[1] p[1],
   d[2] p[1] + d[1] p[2],
   d[3] p[1] + d[2] p[2] + d[1] p[3],
   d[4] p[1] + d[3] p[2] + d[2] p[3],
   d[5] p[1] + d[4] p[2] + d[3] p[3],
   d[5] p[2] + d[4] p[3],
   d[5] p[3]}
```

When you know the impulse response of a system, you can see what effect
the system has on waveforms that are passed through it by convolving the waveform with the impulse response. For example, a system has a top-hat impulse

response, **psfx**, three samples long. What happens when a signal, **datax**, which is the sum of high and low sinusoids, passes through the system? First, you need to generate the lists **psfx** and **datax**; we have plotted **datax** just to see what it looks like before passage through the system:

In:
```
psfx={1,1,1}/3;
datax=Table[Sin[2x]+Sin[x/25],{x,1,50}];
ListPlot[datax,PlotJoined->True,
        AxesLabel->{"sample","value"}]
```

Out:

The convolution lets you see how badly the impulse response of the system affects the components of the input signal:

In:
```
ListPlot[Table[N[Sum[datax[[j]] psfx[[i-j+1]],
                {j,
                Max[1,i-Length[psfx]+1],
                Min[i,Length[datax]],1}]],
            {i,
             1,
             Length[datax]+Length[psfx]-1,
             1}],
         PlotJoined->True,
         AxesLabel->{"sample",
                     "output value"}]
```

Out:

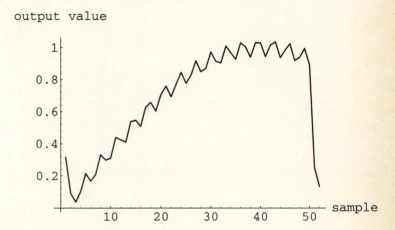

6.1.5 Synthesis of Functions

When you generate time series, you will often define a function or set of functions (which interact in some way) whose results are the values of the time series. Synthesizing time-series functions from other functions is often convenient where it would be difficult to generate the required result directly. Synthesis of functions from others is also a gentle introduction to the foundations of Fourier analysis [Connor82]. To perform the synthesis, you can use **Sum**. You can synthesize a square wave, **ssw**, from a series of cosine functions and then plot the resulting time series by assigning numerical values to the symbolic variables:

In:
```
ssw=Sum[(-1)^((n-1)/2) Cos[w n t]/n,{n,1,11,2}]
```

Out:
$$
Cos[t\ w] - \frac{Cos[3\ t\ w]}{3} + \frac{Cos[5\ t\ w]}{5} - \frac{Cos[7\ t\ w]}{7} +
$$
$$
\frac{Cos[9\ t\ w]}{9} - \frac{Cos[11\ t\ w]}{11}
$$

In:
```
Plot[Normal[ssw] /. {t->x,w->2},{x,0,2 Pi},
    AxesLabel->{"x","value"}]
```

Out:

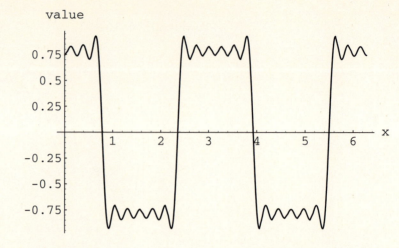

In:
```
ssw1=Sum[(-1)^((n-1)/2) Cos[w n t]/n,{n,1,23,2}];
Plot[Normal[ssw1] /. {t->x,w->2},{x,0,2 Pi},
    AxesLabel->{"x","value"}]
```

Out:

value

In:
```
stw=Sum[(-1)^((n-1)) Sin[w n t]/n,{n,1,20,1}];
Plot[Normal[stw] /. {t->x,w->2},{x,0,2 Pi},
    AxesLabel->{"x","value"}]
```

Out:

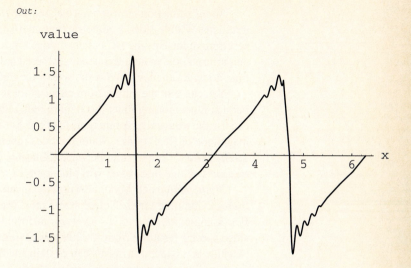

6.2 Fourier Analysis

Mathematica and *Nodal* provide a number of tools with which you can carry out transformation between time and frequency domains. To calculate (complex) frequency spectra, you can use either *Mathematica*'s **Fourier** function or *Nodal*'s **FourierEE** function. The length of the input array to either of these two functions or their inverses is not limited to a power of 2:

In:
```
timeSeries={1,1,1,1,0,0,0,0,1,1,1,1,0,0,0,0};
fTimeSeries=Fourier[timeSeries]
```

Out:
```
{2., 0., 0.5 + 1.20711 I, 0., 0., 0.,
      0.5 + 0.207107 I, 0., 0., 0.,
      0.5 - 0.207107 I, 0., 0., 0.,
      0.5 - 1.20711 I, 0.}
```

In:
```
FourierEE[timeSeries]
```

Out:
```
{1., 0, 0.25 - 0.603553 I, 0, 0, 0,
      0.25 - 0.103553 I, 0}
```

Nodal's function returns a normalized, single-sided spectrum, rather than the traditional double-sided symmetric (or butterfly) spectrum. The first term in each spectrum represents the DC or zero-frequency component; subsequent terms contain information about a frequency f, where $f = 1/nd$, where n is the number of time-domain samples and d is the size of the time step between samples.

You can plot the output from either of these Fourier transforms, but remember that most plotting functions require real, rather than complex, numbers. You can use the **Abs** and **Arg** functions to express complex numbers in terms of (real) magnitude or (real) phase. *Nodal* provides functions such as **ComplexToPolarDegree** for conversions between Cartesian, $\{x, Iy\}$, and other representations. **NodalPlot** and **BodePlot** will plot complex numbers directly.

Mathematica and *Nodal* also provide inverse transforms for converting spectra back into time series. Note that you must use the matching inverse transform. For example, *Mathematica*'s inverse transform will not process correctly single-sided spectra resulting from *Nodal*'s **FourierEE** function. Due to the effects of machine precision, transforms can contain small-valued terms that are not significant, and you can remove them using **Chop**:

```
In:
  InverseFourierEE[%]

Out:
  {1., 1., 1., 1., 0, 0, 0, 0, 1., 1., 1., 1., 0, 0, 0, 0}
```

N

```
In:
  InverseFourier[fTimeSeries]

Out:
                            -18
  {1., 1. - 1.16552 10    I, 1.,

                    -19
    1. - 3.25261 10     I,

              -20
    -5.42101 10     I,

                  -19              -18
    -5.42101 10     + 1.16552 10     I,

                  -19
    -5.42101 10     ,

                  -18             -19
    -1.89735 10     + 3.25261 10     I,
```

$$1., \quad 1. - 1.16552\ 10^{-18}\ \text{I}, \quad 1.,$$

$$1. - 3.25261\ 10^{-19}\ \text{I},$$

$$-5.42101\ 10^{-20}\ \text{I},$$

$$-5.42101\ 10^{-19} + 1.16552\ 10^{-18}\ \text{I},$$

$$-5.42101\ 10^{-19} ,$$

$$-1.89735\ 10^{-18} + 3.25261\ 10^{-19}\ \text{I}\}$$

In:
Chop[%]

Out:
{1., 1., 1., 1., 0, 0, 0, 0, 1., 1., 1., 1., 0, 0, 0, 0}

6.3 Frequency-Domain Filtering

You can use the Fourier principle — that a data series can be expressed as a sum of orthogonal functions — to filter data in the frequency domain. For example, if your data have been corrupted by the addition of noise, the noise may have a different spectral signature from that of the data. In general, knowing either the spectral signature of your data or the corrupting noise enables you to improve the signal-to-noise ratio (SNR) of the data by filtering. The subject of filtering is large; we give an introduction to using *Mathematica* to solve both analog and time-domain digital filtering problems in Chapter 8. Filtering in the frequency domain is tractable for certain filtering tasks.

In the following example, we generate pure data and noise (actually, another sinusoidal signal at a different frequency) and add them together to simulate corrupted data. We Fourier transform the corrupted data to show how the spectral signature of the data and noise can be separated in the frequency domain, and we then apply a crude frequency filter. The filter selects the DC and first three frequency terms in the spectrum. We apply the filter to the complex spectrum by multiplying together the filter and spectrum lists. We then reconstruct the original data from the filtered spectrum using the inverse Fourier transform.

We present all the graphics output in one place, so that you can compare the graphs easily. To prevent each invocation of **ListPlot** generating a plot, we have suppressed production of the plots by setting **DisplayFunction-> Identity**:

```
In:
   pureData=Table[Sin[(2Pi t/16)] //N,
                  {t,0,32}];
   purePlot=ListPlot[pureData,
           PlotJoined->True,
           AxesLabel->{"t","f(t)"},
           PlotLabel->"Pure sine wave",
           DisplayFunction->Identity];


   noise=Table[Sin[(2Pi t/5)]//N,{t,0,32}];
   data=noise+pureData;
   dataPlot=ListPlot[data,
                     PlotJoined->True,
                     AxesLabel->{"t","f(t)"},
                     PlotLabel->"Noisy data",
                     DisplayFunction->Identity];
   complexSpectrum=Fourier[data];
   powerSpectrum=Abs[complexSpectrum];
   powerPlot=ListPlot[powerSpectrum,
                     PlotJoined->True,
                     PlotRange->All,
                     AxesOrigin->{0,0},
                     AxesLabel->{"frequency\nbin",
                                 "power"},
                     PlotLabel->"Spectrum",
                     DisplayFunction->Identity];


   filter=RotateLeft[Table[If[i<7, 1, 0],
                     {i,0,32}],
                  3]
```

```
Out:
   {1, 1, 1, 1, 0, 0, 0, 0, 0, 0, 0, 0, 0, 0, 0, 0, 0, 0,
   0, 0, 0, 0, 0, 0, 0, 0, 0, 0, 0, 1, 1, 1}
```

```
In:
   filtPlot=ListPlot[Re[InverseFourier[
                        filter complexSpectrum]],
                  PlotLabel->"Filtered data",
                  PlotRange->All,
                  AxesLabel->{"t","f(t)"},
                  PlotJoined->True,
```

```
                              DisplayFunction->Identity];
                Show[GraphicsArray[{{purePlot,dataPlot},
                                    {powerPlot,filtPlot}}]]
```

Out:

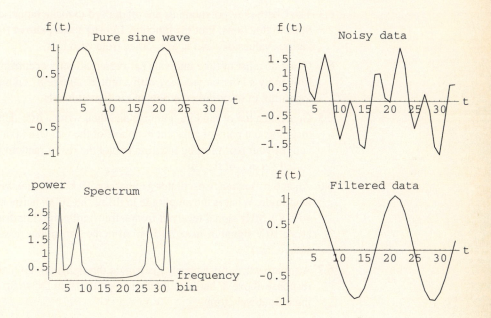

6.4 Summary

In this chapter you have seen how

- to generate time series using *Mathematica*
- to correlate different time series to establish when they are similar
- to convolve a time series with an impulse response to mimic instrument responses
- to synthesize waveforms from functions at different frequencies and amplitudes
- to perform Fourier analysis and elementary frequency-domain filtering.

6.5 Exercises

6.1 Generate a short, simple time series and then calculate the differences between successive members of the series.

6.2 How would you sort or reverse the order of a time series?

6.3 The Chebyshev polynomials are often used to make numerical approximations to functions. Plot the function $1/(1+ (r\ \mathtt{ChebyshevT}[n,\ x])^2)$ for various values of r and n, as x varies from zero to π.

6.4 Generate a list of time intervals between randomly occurring events (driven by a Poisson process) with a unit mean rate by taking the natural logarithm of uniformly distributed random numbers.

6.5 Generate a list of random numbers and then write a user-defined function that plots out pairs of adjacent members of the list. Use (a) a bin-count plot and (b) your pair-plotting function to plot the list of random numbers and a sorted version of the list.

6.6 An audio-frequency oscillator generates a 10 kHz square wave of 1 V amplitude. What is the amplitude in dB of the 100th harmonic compared with a 1 MHz signal received from a transmitter 100 km distant that creates a 10 V signal in an aerial 10 m from the transmitter (assume an inverse square law)?

6.7 A square wave of frequency f is passed through a circuit that has no response above $8f$ but has a perfect response from DC to $8f$. Plot the resulting output waveform.

6.6 References

[Connor82] Connor, F.R., *Signals*, Edward Arnold, London, 1982.

[DeFatta88] DeFatta, D.J., Lucas, J.G., Hodgkiss, W.S., *Digital Signal Processing: A System Design Approach*, John Wiley & Sons, New York, 1988.

[Lynn82] Lynn, P. A., *An Introduction to the Analysis and Processing of Signals*, Macmillan Press, London, 1982.

CHAPTER 7

s-Domain (Laplace) Analysis

Laplace circuit analysis can describe the behavior of a circuit in both the time and the frequency domain. Because you can determine analytically the *s*-domain (Laplace) transfer function of a circuit relatively easily, Laplace analysis is a commonly used, important technique.

7.1 Laplacian Description of Signals

When you analyze a signal using Fourier techniques, the result is a spectrum. A particular point in the spectrum represents the contribution of a sinusoidal frequency of a particular frequency and, because the spectrum is complex, that particular frequency's phase. The contribution at that particular frequency is assumed to be of constant amplitude (and phase). For many signals in the real world, the constancy of the contribution is inappropriate: Signals often change amplitude and may not be sinusoidal.

The Laplacian description is more general than the Fourier. A signal has two descriptive attributes: One describes the frequency of the signal, and the other describes the extent to which the signal, which may have no sinusoidal component, is exponentially increasing or decaying. These two attributes, ω and σ, respectively, are mapped as coordinates onto a Cartesian plane called the *s*-plane, where $s = \sigma + I\omega$ [Lynn82].

You can use *Mathematica* to visualize the form of signals with different values of ω and σ. Here is a function, **sPlot**, that takes numeric values for ω and σ and plots $E^{(\sigma + I\omega)t}$ on the complex plane and shows the real and imaginary components. You can use **sPlot** to visualize waveforms that have, for example, both frequency and exponential terms, or that have only a frequency term that is a function of time:

```
In:
  sPlot[sp_,w_]:=
    Block[{zt1,ztlist1,ztylist1},
          zt1=Table[N[Exp[(sp+I w)t]],
                     {t,0,Pi,Pi/100}];
          Show[GraphicsArray[
            {ListPlot[Map[({Re[##],Im[##]})&, zt1],
```

```
                    PlotLabel->"Complex plane",
                    AspectRatio->1,
                    DisplayFunction->Identity],
                ListPlot[Map[Re[##]&, zt1],
                    PlotLabel->"Real component",
                    PlotJoined->True,
                    DisplayFunction->Identity],
                ListPlot[Map[Im[##]&, zt1],
                    PlotLabel-> "Imaginary component",
                    PlotJoined->True,
                    DisplayFunction->Identity]}]]]
```

sPlot[1,10]

Out:

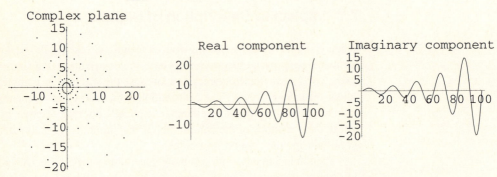

Chirp and sweep waveforms have a frequency that is a function of time. Here is a constant-amplitude sinusoidal sweep waveform:

In:
sPlot[0, 10t]

Out:

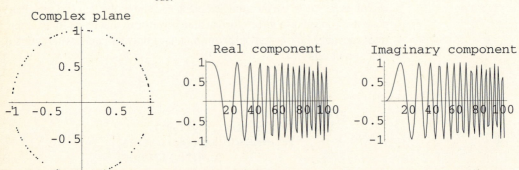

7.2 *s*-Domain Transfer Functions

You have calculated the frequency-domain transfer function of circuits in Chapters 3 and 4: The transfer function, or gain, is computed at a specified frequency. Similarly, the response of the circuit to an *s*-domain signal is described by the circuit's *s*-domain transfer function. You can compute the *s*-domain transfer function from the *s*-domain nodal admittance matrix (NAM), which you can construct using the method we described in Chapter 4, except that entries in the NAM for resistors, capacitors, and inductors are $1/R$, sC, and $1/sL$, respectively [Adby80].

If you set the **Frequency->** rule to **Laplace** in the *Nodal* function **NodalAnalyze**, it will compute the *s*-domain transfer function of a network for you, directly from the circuit's netlist. For example, we can define a simple RC integrator circuit, **filter**, and then determine its *s*-domain forward transfer function in terms of ABCD-parameters. We then use the **NodalAnalyze** command to determine an expression for the output voltage, **v5**, from a Sallen–Key filter:

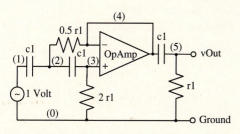

```
In:
  Needs["Nodal`","Nodal2.m"];
  filter=NodalNetwork[Resistor[{1,2},r1],
                      Capacitor[{2,0},c1]];
  ABCDParameters[filter,
                 Frequency->Laplace]

Out:
     1+c1 r1 s         r1
     c1 s              1
```

```
In:
  rcFilterTF=1/Normal[%][[1,1]]

Out:
         1
     -----------
     1 + c1 r1 s
```

```
In:
  SKfilter=NodalNetwork[
               VoltageSource[{1,0},1],
               Capacitor[{1,2},c1],
               Capacitor[{2,3},c1],
               OpAmp[{3,4,4}],
               Resistor[{3,0},2 r1],
               Capacitor[{4,5},c1],
               Resistor[{2,4},0.5 r1],
               Resistor[{5,0},r1]];
```

```
NodalAnalyze[SKfilter,
             Result->V5,
             Frequency->Laplace]
```

Out:

$$
\cfrac{c1^3 \ s^3}{\cfrac{1.}{r1^3} + \cfrac{2. \ c1 \ s^2}{r1^2} + \cfrac{2. \ c1^2 \ s^2}{r1} + c1^3 \ s^3}
$$

7.3 Visualization of Pole-Zero Descriptions

You can use *Mathematica*'s plotting facilities to help visualize pole-zero descriptions of *s*-domain transfer functions [Connor86]. Visualization is an important aid for studying analog filter design (see Chapter 8), because the magnitude of a circuit's transfer function for a given value of *s*, TF(*s*), indicates how a signal with its Laplace transform equal to that value of *s* will be modified as that signal passes through the circuit: If a zero exists at s_1 — that is $TF(s_1) = 0$ — then signals with $s = s_1$ will be completely attenuated.

In the next example, we determine the *s*-domain transfer function for a simple RLC filter, determine analytically where the transfer function becomes infinite (that is, where its poles are located), and then plot out the surface of the transfer function as a function of σ and ω.

In:
```
net=NodalNetwork[VoltageSource[{1,0},vIn],
                 Resistor[{1,2},r1],
                 Inductor[{2,3},h1],
                 Capacitor[{3,0},c1]];
Simplify[NodalAnalyze[net, Result->V3/V1,
                      Frequency->Laplace]]
```

Out:

$$
\frac{1}{1 + c1 \ r1 \ s + c1 \ h1 \ s^2}
$$

You can substitute the resonant frequency and selectivity factor, **w0** and **q**, for combinations of R, L, and C. This substitution parameterizes the circuit in terms of more useful variables. You can find where the transfer function becomes infinite by determining the values of s ($= \sigma + I\omega$) that make the denominator zero, thus identifying the pole locations:

```
In:
  % //. {h1 c1->(1/w0)^2, r1->w0 h1/q}
```

```
Out:
          1
    --------------
          2
        s       s
  1 + --- + ----
          2   q w0
        w0
```

```
In:
  tf=Normal[%][[1,2]];
  Solve[Denominator[tf]==0,s]
```

```
Out:
              w0    Sqrt[1 - 2 q] Sqrt[1 + 2 q] w0
           -(--) + ------------------------------
              q                   q
  {{s -> ---------------------------------------},
                          2


              w0    Sqrt[1 - 2 q] Sqrt[1 + 2 q] w0
           -(--) - ------------------------------
              q                   q
   {s -> ---------------------------------------}}
                          2
```

Defining the transfer function for specific values of **q** and **w0** (**q->2** and **w0->10**) enables you to visualize the poles in the transfer function:

```
In:
  transferFunc[s1_Complex]:=tf /.
                        {s->s1,q->2,w0->10};
  Plot3D[Abs[transferFunc[sigma + omega I]],
        {sigma,-7,8},{omega,-15,15},
          AxesLabel->{"s","w","tf"},PlotPoints->59,
          PlotRange->{{-7,8},{-15,15},{0,25}},
          Shading->False, BoxRatios->{1,1,1}]
```

Out:

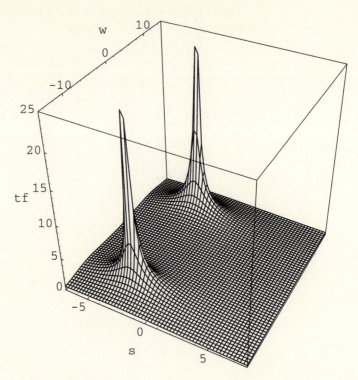

7.4 Determination of Circuit Impulse and Step Responses

By taking the inverse Laplace transform of the transfer function (equal to the reciprocal of A_{11}, in our circuit **filter**), you can determine the circuit's impulse response. By integrating the impulse response with respect to **t**, from zero to **t1**, you can calculate the time-domain response of the circuit to a unit voltage step at its input.

Toolbox

Laplace Transform

You load the Laplace transform package by typing
`<<Calculus`LaplaceTransform`.`

LaplaceTransform[*expr*,*t*,*s*] calculates the Laplace transform of *expr* as a function of *t*.

The inverse operation is performed by
InverseLaplaceTransform[*expr*,*s*,*t*].

Integration

Integrate[*expr*,*var*] gives the indefinite integral of *expr* with respect to *var*.

Integrate[*expr*, {*var*, *min*, *max*}] gives the definite integral over the range from *min* to *max*.

```
In:
   InverseLaplaceTransform[rcFilterTF,s,t]
Out:
                      1
        ------------------------
                 t/(c1 r1)
        c1 r1 E


In:
   Integrate[%,{t,0,t1}]
Out:

        -(t1/(c1 r1))
   1 - E
```

The result of applying a voltage step to circuit **filter** is an exponentially rising voltage at the output, with the rate of rise being controlled by the values of the capacitor and resistor. Using a particular voltage level as a trigger point allows you to use this type of circuit in relaxation oscillators.

As another example of manipulating circuit analysis in the *s*-domain, the next example shows how a (relatively) complex problem can be solved with help from *Mathematica* (with considerable time savings compared to the equivalent manually executed process).

First, we define a bridged-T filter in which inductors are assumed equal and are assigned the values **h1** (in henries). Then we analyze the circuit to obtain the voltage gain using the Y-parameter expression —**Y21/Y22** — which is equivalent to the voltage gain under the ideal conditions of zero voltage-source impedance and infinite load impedance:

In:
```
bt=NodalNetwork[Inductor[{1,2},h1],
                Inductor[{2,3},h1],
                Resistor[{1,3},r1],
                Capacitor[{2,0},c1]];
NodalAnalyze[bt,
             Nodes->{1,3},
             Result->-Y21/Y22,
             Frequency->Laplace]
```

Out:
```
     1            2         c1 s
  ------  +  -------  +  ----
    2   2     h1 r1 s     r1
  h1   s

  ---------------------------
  c1      1           2        c1 s
  --  +  ------  +  -------  +  ----
  h1      2   2     h1 r1 s      r1
        h1   s
```

You can display this expression in a simpler form by using **Simplify**, before assigning unit values to all the components:

In:
```
Simplify[Normal[%][[1,2]]]
```

Out:
```
                              2   3
           r1 + 2 h1 s + c1 h1   s
  -------------------------------------
                            2           2  3
  r1 + 2 h1 s + c1 h1 r1 s   + c1 h1   s
```

In:
```
tf = % /. {r1->1, h1->1, c1->1}
```

Out:
```
               3
  1 + 2 s + s
  -----------------
              2    3
  1 + 2 s + s  + s
```

In this example, to find the time-domain response of the bridged-T circuit to a unit step, we multiply the *s*-domain transfer **tf** by 1/*s*, the unit step's function in the *s*-domain, and then take the inverse Laplace transform of the result:

```
In:
  %/s
```

```
Out:
                      3
        1 + 2 s + s
      --------------------
                    2    3
      s (1 + 2 s + s  + s )
```

In theory, you should be able to take the inverse Laplace transform of this equation to obtain the time-domain response of the circuit. In practice, it is difficult to recognize the various components contained in this somewhat condensed expression and that make the inverse transform task difficult. To make the task of inverse transformation more tractable, you can use *Mathematica* to help you in splitting up the condensed form into one that is more open and that *Mathematica* is then able to invert. *Mathematica* is often able to factor polynomials, but the factoring works with only integer or exact rational coefficients. In this example, we show you how to split up the denominator. There is nothing special about the method we used here to split up the condensed form of the equation. It is just one way that happens to work and that follows general guidelines about splitting any complicated expression into partial fractions.

To factor the denominator, we can start by arbitrarily decomposing `s(1+2s+s^2+s^3)` into `s(s+a)(s+b)(s+d)`, thus leaving the task of finding values for **a**, **b** and **d**; we have not used the variable **c** because *Nodal* considers **c** to be equated to the speed of light. You can establish a set of equations, the solutions to which give **a**, **b**, and **d** (which you can find using **Solve**), by expanding `s(s+a)(s+b)(s+d)` and comparing terms with those of `s(1+2s+s^2+s^3)` of equal power in **s**. You can simplify the task of comparing the coefficients by using *Mathematica*'s **CoefficientList** function to generate a list of coefficients (**coeffs**) and then solving numerically for equality with the numerical values, found by inspection:

```
In:
  result=%;
  try=s(s+a)(s+b)(s+d);
  Expand[try];
  coeffs=CoefficientList[%,s]
```

```
Out:
  {0, a b d, a b + a d + b d, a + b + d, 1}
```

```
In:
  abdValues=
  Chop[NSolve[coeffs=={0,1,2,1,1},{a,b,d}]]
```

```
Out:
  {{a -> 0.21508 + 1.30714 I, b -> 0.56984,
      d -> 0.21508 - 1.30714 I},
   {a -> 0.56984, b -> 0.21508 + 1.30714 I,
      d -> 0.21508 - 1.30714 I},
   {a -> 0.21508 - 1.30714 I, b -> 0.56984,
      d -> 0.21508 + 1.30714 I},
   {a -> 0.56984, b -> 0.21508 - 1.30714 I,
      d -> 0.21508 + 1.30714 I},
   {a -> 0.21508 + 1.30714 I, b -> 0.21508 - 1.30714 I,
      d -> 0.56984},
   {a -> 0.21508 - 1.30714 I, b -> 0.21508 + 1.30714 I,
      d -> 0.56984}}
```

Now that we know the coefficient values, we can split up the product of the transfer and unit-step functions, **result**, into simple enough partial fractions that *Mathematica* can compute the inverse Laplace transform of the partial fractions.

```
In:
  trySplit=N[Chop[Apart[Numerator[result]/
              (try //. abdValues[[1]])]],2]

Out:
  1.    -0.16 + 0.34 I      -0.16 - 0.34 I       0.31
  -- + ---------------- + ---------------- + --------
  s    0.22 - 1.3 I + s   0.22 + 1.3 I + s   0.57 + s
```

You can confirm that the expanded form is equivalent to the equation with which you started by gathering together the equation terms over a common, expanded denominator.

Toolbox

Manipulating Expressions

Together[*expr*] collates the terms in *expr* over a common denominator and cancels common factors.

Apart[*expr*] reformats *expr* as a sum of terms with simple denominators.

Expand[*expr*] expands products and positive integer powers in *expr*.

ExpandAll expands out all products and powers.

ExpandNumerator and **ExpandDenominator** manipulate specific parts of an expression.

In:
```
Together[trySplit];
Chop[ExpandDenominator[%]]
```

Out:

$$\frac{1. + 2. \; s + 1. \; s^3}{1. \; s^2 + 2. \; s^3 + 1. \; s^4 + s}$$

You can now obtain the circuit's response to a step input by taking the inverse Laplace transform of **trySplit** (the product of the *s*-domain transfer function and the Laplace transform for a step, 1/*s*) and then plot the output voltage resulting from the application of a unit step voltage to the circuit's input, as a function of time. (We have used **Chop** in the **Plot** command to excise small-valued complex components that result from our use of low-precision real numbers.)

In:
```
outputVoltage=InverseLaplaceTransform[trySp,t]
```

Out:

$$1. + \frac{0.31}{E^{0.56984t}} + (-0.16 - 0.34 \; I) E^{(-0.21508 - 1.30714 \; I)t} +$$

$$(-0.16 + 0.34 \; I) E^{(-0.21508 + 1.30714 \; I)t}$$

In:
```
Plot[outputVoltage//Chop, {t, 0, 5Pi},
  PlotRange->{0,2},AxesLabel->{"time","voltage"}]
```

Out:

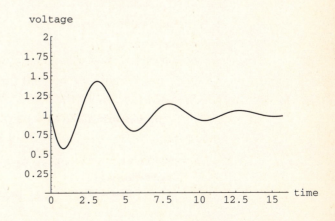

The steady-state *s*-domain transfer function gives the transfer function of the circuit when all the transients due to circuit start-up have died away. You can calculate the steady-state transfer function in the frequency domain by substituting **s = I w** into the *s*-domain transfer function:

In:

```
tf /. s->(w I)
```

Out:

$$
\frac{1 + 2 I w - I w^3}{1 + 2 I w - w^2 - I w^3}
$$

Our last example is another T-format circuit. This time, we calculate both frequency- and time-domain characteristics to show both how they relate and how they differ:

In:

```
tbridge=NodalNetwork[VoltageSource[{1,0},vIn],
                     Resistor[{1,2},r1],
                     Inductor[{2,3},h1],
                     Capacitor[{3,0},c1],
                     Inductor[{3,4},h1],
                     Resistor[{4,0},r2]];
tbtf=NodalAnalyze[tbridge,
                  Result->V4/V1,
                  Frequency->Laplace]
```

Out:

$$
-\left(1 \,/\, \left(h1^2\ r1^2\ s^2\ \left(-\left(\frac{c1}{h1\ r1}\right) - \frac{c1}{h1\ r2} - \frac{1}{h1^2\ r1^2\ s^2}\right.\right.\right. \\
\left.\left.\left. - \frac{1}{h1^2\ r2^2\ s^2} - \frac{c1}{h1^2\ s^2} - \frac{2}{h1\ r1\ r2\ s} - \frac{c1\ s}{r1\ r2}\right)\right)\right)
$$

In:

```
parameterRules={h1->q r2/(2 Pi),
                c1->1/(q r2 2 Pi),
                r1->r2,
                s->2 Pi I f};
pTF=Normal[
          Simplify[tbtf /. parameterRules]][[1,2]]
```

Out:

$$\frac{q}{I\,f + 2\,q - 2\,f^2\,q + 2\,I\,f\,q^2 - I\,f^3\,q^2}$$

To visualize the frequency-domain transfer function, we can plot it as a function of frequency for three values of **q**. The function **Plot** can take a list of multiple functions as its first argument. We create a list of three functions by using **Evaluate** to force **Plot** to evaluate its first argument (so creating a list of three functions) prior to substituting numerical values of **f**:

In:
```
Plot[Evaluate[Abs[pTF] /. q->{0.1,1,10}],
     {f,0,5},
     AxesLabel->{"f","tf"}]
```

Out:

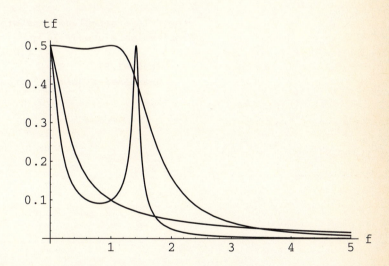

For convenience we create the step response by dividing the transfer function, **ltf**, by **p** before taking the inverse Laplace transform — which corresponds to integrating the time-domain response, so converting the impulse response into a step response.

By inspection, the first two terms of the step response imply an exponential increase in the output signal. The latter terms modify the rise time and create ringing in the waveform through an exponentially decaying sinusoid. The circuit's behavior is clearer if we plot the responses for a few values of **qo**.

Here, the variables **qo** are scaled by **2 Pi** so they correspond to the **q** used in the previous frequency-domain plots. The fastest rise time is given by the circuit with the lowest **q** because that circuit has the greatest high-frequency content. The high-**q** network may look unusable because of its time-domain response.

```
In:
  ltf=
      Factor[Simplify[pTF /.
                           {f->p/(2Pi I),
                            q->2 Pi qo}]]
```

```
Out:
                        2
                4 Pi  qo
       ----------------------------------
                        2        2
       (1 + p qo) (p + p  qo + 8 Pi  qo)
```

```
In:
  stepTF=InverseLaplaceTransform[ltf/p, p, t]
  Plot[Evaluate[stepTF /. qo->
                          ({0.1,1,10}/(2Pi))],
       {t,0.01,5},
       AxesLabel->{"time","output voltage"}]
```

```
Out:
  1       1
  - - ------- - (-(
  2       t/qo
      2 E

                                        2    2
        -t/(2 qo) - (Sqrt[1 - 32 Pi  qo ] t)/(2 qo)
       E
       -------------------------------------------) \
                               2    2
                   Sqrt[1 - 32 Pi  qo ]

                                        2    2
        -t/(2 qo) + (Sqrt[1 - 32 Pi  qo ] t)/(2 qo)
       E
     + -------------------------------------------) \
                               2    2
                   Sqrt[1 - 32 Pi  qo ]


       / 2
```

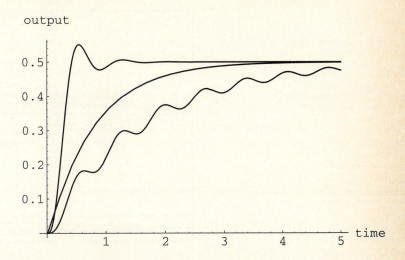

You can use *Mathematica* to hear what these waveforms sound like, if your computer supports the sound facilities in *Mathematica*. The frequency of the q=10 oscillatory waveform is rather low to hear, but you can use it to amplitude-modulate a higher-frequency waveform, or carrier. First, you need to generate two lists that contain the q=10 waveform and a 200 Hz carrier. Each list contains 3 s of millisecond-spaced values.

In:

```
lowQ=Table[Chop[N[stepTF /. qo->10,3]],
          {t,0,3,0.001}];
carrier=Table[Sin[2 Pi 200 t],
             {t,0,3,0.001}];
```

You create the amplitude-modulated sound, **amSound**, by multiplying together the carrier and the low-frequency signal and by invoking the function **ListPlay**:

In:

```
amSound=carrier lowQ;
ListPlay[amSound,
        SampleRate- >1000,
        PlayRange->All]
```

Out:

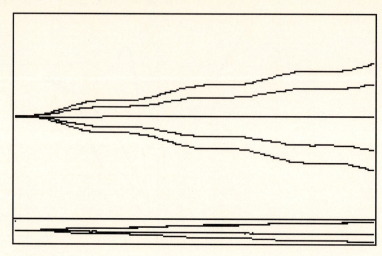

–Sound–

7.5 Summary

In this chapter, you have seen how

- to visualize signals with different *s*-plane descriptions
- to calculate *s*-plane transfer functions
- to visualize *s*-plane transfer functions with poles and zeros
- to derive the frequency- and time-domain behavior of circuits from *s*-domain analysis.

7.6 Exercises

7.1 Plot signals with *s*-plane coordinates {1, 1}, {−1, 1}, {−1, −1}, {1, −1}, {1, 0}, {0, 1}.

7.2 Find the zeros and poles of the transfer function $(s^3 + 4s)/(s^2 + 1)$.

7.3 Use **Plot3D** to display the transfer function $(s^3 + 4s)/(s^2 + 1)$.

7.4 Determine the step response of the two circuits shown in Figure 7.1.

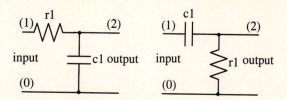

Figure 7.1   Integrator and differentiator circuits.

7.5 Determine the step response of a potential divider circuit that uses capacitors instead of resistors. What happens to the output voltage in practice?

7.7 References

[Adby80] Adby, P. R., *Applied Circuit Theory*, Ellis Horwood, Chichester, United Kingdom, 1980.

[Connor86] Connor, F.R., *Networks*, Edward Arnold, London, 1986.

[Lynn82] Lynn, P. A., *An Introduction to the Analysis and Processing of Signals*, Macmillan Press, London, 1982.

CHAPTER 8

Filter Design

There are two complementary goals in electronic design: to change the amplitude of a signal without distortion and to distort a signal in a specified manner. Creating a desired frequency content in a signal, by filtering it, is distortion raised to the level of an art. Nowhere in electronics does abstract mathematics merge with engineering practicality with greater elegance. You will find *Mathematica* very useful in working with the mathematics of filters. The ease of plotting, constructing transfer functions, and extracting pole locations will make easy a theoretically intimidating and algebraically tedious subject.

8.1 Transfer Functions

Standard filter transfer functions, Butterworth and Chebyshev, for example, have been developed as a result of years of work in design, and their implementations have useful frequency responses and easily calculated component values [Kuo66]. Where the component values are not easily calculated, extensive tables have been prepared [Zverev67]. Tables and formulae do not help you to visualize a transfer function — but *Mathematica* can make visualization easier for you.

The simplest form of transfer function is often referred to as a brick wall: On one side, the signal is unmodified; on the other side, the signal is infinitely attenuated. Here is what a brick-wall transfer function looks like:

```
In:
 Plot[  If[ f <= 1, 1, 0 ],
         {f,0,3},
         AxesLabel->{"Hz", "Mag"},
         PlotLabel->
           "Brick-wall transfer function"];
```

Out:

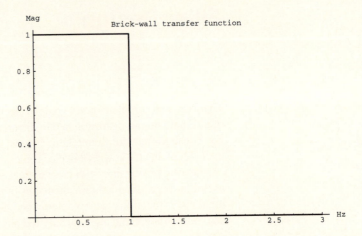

The brick-wall shape is an idealized representation of a low-pass filter's frequency-domain transfer function, but it is unrealizable because its impulse response is a **Sinc** function, sin(*t*)/*t*, that is infinite in extent, thus implying an infinitely large number of components. To show the impulse response, we define a **Sinc** function in *Mathematica*. Just defining **Sinc[t]** would, however, be unsatisfactory because of the singularity at **t** = 0. We can avoid this potential problem by defining another version of the function that checks for its argument being both numerical and within 10^{-9} of the origin. Both definitions coexist: The function is overloaded. We use a condition rule (**/;**) and a test for a numerical argument (**?NumberQ**) to avoid the singularity at the origin:

In:

```
Sinc[t_] := Sin[Pi t]/(Pi t)
Sinc[t_?NumberQ] := 1 /; Abs[t] < 10^-9


Plot[   2 Sinc[2 t],
        {t,-3,3},
        AxesLabel->{"Sec","Ampl"},
        PlotLabel->"Brick-wall impulse response",
        PlotRange->All];
```

Out:

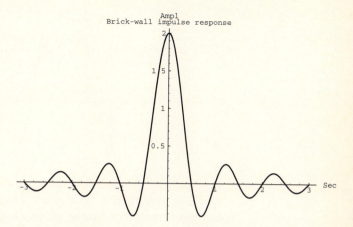

Butterworth and Chebyshev filters are realizable approximations to the brick-wall filter. You can also design filters via time-domain properties, such as minimal ringing (Gaussian) or minimal delay distortion (Bessel) — there are many design criteria. In Sections 8.1.1 through 8.1.4, we describe the frequency response of Butterworth and Chebyshev filters, their *s*-plane surfaces, and the calculation of component values.

8.1.1 The Butterworth Response

The simplest realizable transfer function, which has a maximally flat response, is the Butterworth. Butterworth filters provide a monotonic attenuation (roll-off) with frequency. As the order of a Butterworth filter is increased to infinity, the response approaches a brick-wall filter shape. The product of left- and right-plane pole locations, $H(s)\,H(-s)$, for a *n*th order filter has the simple form $1/(1 - s^{2n})$, which we have defined as the function **h2B**. Once we let *s* become $I\omega$ (that is, we are dealing with steady-state sinusoidal signals), then **h2B** equals the magnitude squared of our transfer function. We can synthesize filters directly, by using **h2B**:

```
In:
  h2B[n_,s_] := 1/(1 + (-s^2)^n)

  Plot[ Release[Sqrt[Abs[h2B[#,I w]& /@ {1,3,10}]]],
      {w,0,3},
      AxesLabel->{"rad/s","Mag"},
      PlotLabel->"Butterworth transfer functions",
      PlotRange->All];
```

Out:

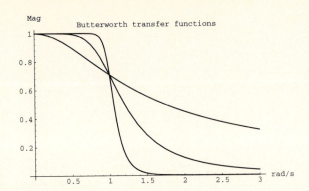

Each response is $1/\sqrt{2}$ at $\omega = 1$, which makes a frequency of 1 rad/s the half-power point. You have to use very high filter orders to make a Butterworth filter perform close to a brick-wall transfer function.

8.1.2 The Chebyshev Response

The Chebyshev response is a better approximation to a brick-wall filter than is the Butterworth design; the improvement is achieved by allowing ripples in the passband. The Chebyshev transfer function is actually a set of polynomials defined, for orders 0, 1, and *n*, by a recursion formula:

In:
```
ChebyP[0, w_]:= 1
ChebyP[1, w_]:= w
ChebyP[n_,w_]:= 2 w ChebyP[n-1,w] - ChebyP[n-2,w]
```

(*Mathematica* has its own functions for Chebyshev polynomials, but **ChebyP** demonstrates how you can easily declare your own functions by using overloading and recursion, where appropriate.) So, for example, a fourth-order filter's transfer function becomes

In:
```
ChebyP[4,w]
```
Out:
$$1 - 2 w^2 + 2 w (-w + 2 w (-1 + 2 w^2))$$

In addition to the frequency and filter order variables, **w** and **n**, the Chebyshev transfer function uses a ripple depth parameter, ε (**e** in our *Mathematica* code). Again, we derive the squared transfer function from the product of left- and right-plane pole locations:

In:

```
h2C[n_,s_,e_] := 1/(1 + e^2 ChebyP[n,s/I]^2);

Plot[
  Release[
      Sqrt[Abs[h2C[#,I w,0.4]& /@ {1,3,10}]]],
  {w,0,3},
  AxesLabel->{"Hz","Mag"},
  PlotLabel->"Chebyshev transfer functions",
  PlotRange->All];
```

Out:

Chebyshev functions all have a value of unity at $\omega = 1$. The factor of ε in the Chebyshev transfer function forces the magnitude to be $(1/(1 + \oplus))^{1/2}$ at $\omega = 1$, a frequency often far from the half-power point. Higher ripple magnitudes and more ripples create a steeper roll-off slope in the Chebyshev filter.

8.1.3 Pole-Zero Locations

You can get a tremendous insight into filter behavior from knowing the transfer function pole and zero locations. You can find their locations using *Mathematica*'s **Solve** function. For example, for the Butterworth filter:

In:

```
poles = Solve[1/h2B[3,s] == 0, s]
```

Out:

$$\{\{s \to 1\}, \{s \to (-1)^{1/3}\}, \{s \to (-1)^{2/3}\}, \{s \to -1\},$$
$$\{s \to (-1)^{4/3}\}, \{s \to (-1)^{5/3}\}\}$$

By inspection, we can see that the Butterworth filter's poles lie on the unit circle, and we can extract the root values by mapping the substitution function, `/.`, over the list of poles. Because we only want left-half plane roots, we use **Select** to keep only those values that have real parts less than zero:

In:
```
locs = Map[(s /. #)&, poles]
```

Out:

$$\{1,\ (-1)^{1/3},\ (-1)^{2/3},\ -1,\ (-1)^{4/3},\ (-1)^{5/3}\}$$

In:
```
lhpLoc = Select[locs, (Re[N[#]] < 0)&]
```

Out:

$$\{(-1)^{2/3},\ -1,\ (-1)^{4/3}\}$$

You can plot the poles by using **N** to force a complex number and **Re** and **Im** to extract the real and imaginary parts:

In:
```
ListPlot[Map[{Re[#],Im[#]}&, N[lhpLoc] ],
    AxesLabel->{"sigma","w"},AxesOrigin->{0,0},
    PlotLabel->"Pole Locations",
    PlotStyle->PointSize[0.04],
    PlotRange->{{-1.1,0.1},{-1.1,1.1}},
    AspectRatio->Automatic];
```

Out:

For future use, we combine these functions into a single function called
PolePlot:

In:

```
PolePlot[fcn_,var_] :=
        Module[{poles,locs,lhpLoc},
                poles = Solve[1/fcn == 0, var];
                locs = Map[(var /. #)&, poles];
                lhpLoc = Select[locs,
                            (Re[N[#]] < 0)&];
                ListPlot[Map[{Re[#],Im[#]}&,
                        N[lhpLoc]],
                    AxesLabel->{"sigma","w"},
                    PlotLabel->"Pole Locations",
                    PlotStyle->PointSize[0.04],
                    AxesOrigin->{0,0},
                    PlotRange->{{-1.1,0.1},
                            {-1.1,1.1}},
                    AspectRatio->Automatic
                ]
        ]
```

Solve for the function
poles.
Extract just the numeri-
cal pole location.
Keep only those poles in
the l.h.p.

Plot the poles with
points. Use the real and
imaginary parts of the
pole location as coordi-
nates. Many options to
ListPlot are used to
define the plot.

In:

```
PolePlot[h2C[3,s,0.4], s];
```

Out:

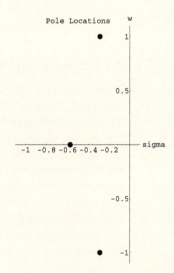

Insight into filter responses comes from observing how pole locations change with filter types and order, so you need a tool to plot them. You can visualize a surface in the *s*-plane (formed by poles and zeros) as a stretched rubber sheet with on-end, upright pencils under the sheet where the poles are located, and tacks to hold the sheet down where the zeros are located. Plotting *s*-plane surfaces in *Mathematica* is easy. Using **Plot3D**, you can see a slice through the rubber sheet along the Iω axis: The level of the sheet along the Iω axis is the magnitude of the filter's transfer function. The resulting plot shows how the poles interact to give a maximally flat response at $\omega\,(=\mathbf{w})=0$:

In:
```
Plot3D[ Sqrt[Abs[h2B[3,sig + I w]]],
        {sig,-1.5,-0.01},{w,-2,2},
        PlotPoints->30,
        ViewPoint->{2.0,-2.5,1}];
```

Out:

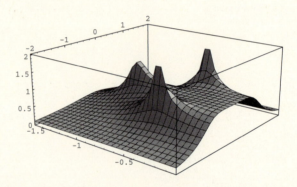

8.1.4 Component Values

Filter component values are determined by pole and zero locations. For Butterworth and Chebyshev filters, explicit formulae exist for the component values. The component-value formulae create filters normalized to $\omega=1$ and $Z_0=1$. In this section, we show how to make a function that returns filter-component values. In Section 8.2, we shall make a function that denormalizes and transforms the filter components, customizing filters to other frequencies and terminating impedances. In keeping with a functional and object-oriented style, we shall use data types (with names such as **Butterworth** and **Chebyshev**) and then request the component values for these data types.

Considering the components used in the filters, you can see that the first and last components are the termination resistances with alternate L- or C-valued components between terminations, as shown in Figure 8.1. The component-value formulae come from Baher's book [Baher84].

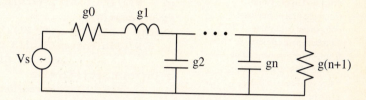

Figure 8.1   Prototype filter.

Although we have focused on passive filters, we have not neglected active filters, because most active filters can be designed directly from the LC prototype values that we shall derive in this section [Matthaei80, Baher84, Rhodes76, Jacobs78].

Recognize only a Butterworth filter type in this function call.

```
In:
Components[Butterworth[n_]] :=
    Module[{thetas},
        thetas = N[(2 Range[n] - 1) Pi/(2 n)];
        Join[ {1}, 2 Sin[thetas], {1}]
    ]
```

The **Components** functions for Butterworth and Chebyshev filters show how easy it is with *Mathematica* to implement the formulae that prescribe filter component values. The **Butterworth** formula is simply a sine function evaluated over a range of angles. The list of Butterworth component values is joined with termination resistor values of 1 Ω.

Two subtleties arise when we use **Range** to create a list of numbers and use **FoldList** to implement the **Chebyshev** function. Because an nth-order Chebyshev formula for an nth-order filter is defined recursively, each additional component requires the last component value. We could use a loop to compute the new value from the last value, but loops are inefficient in *Mathematica*. However, the **FoldList** function does just what we want, because it repeatedly applies a function to a list and to the last result. The only other complication with the Chebyshev filter component values is that the value of the termination resistor depends on the ripple value for even-order networks. Even-order Chebyshev transfer functions do not have unity gain at $\omega = 0$, so the reduced gain must be accounted for in an impedance mismatch.

Recognize only Chebyshev filter specifications.
Get pole eccentricity from order and ripple value.
Compute first normalized component (g1) value.
Compute pole angles.
Compute component ratios.
Use FoldList to compute values and Join to complete list.
Termination value depends on ripple.

```
In:
Components[Chebyshev[n_,e_]] :=
  Module[ {eta,g1,gs,rg},
    eta = N[ Sinh[1/n ArcSinh[1/e]] ];
    g1 = N[2/eta Sin[Pi/(2 n)] ];
    rg = Range[n];
    thetas = N[(2 rg - 1) Pi/(2 n) ];
    gs = N[4Sin[thetas]Sin[(2 rg + 1) Pi/(2 n)]/
                      (eta^2 + Sin[rg Pi/n]^2) ];
    Join[FoldList[(#2/#1)&,1,
                      Prepend[Drop[gs,-1],g1]],
              {If[ EvenQ[n],(e + Sqrt[1+e^2])^2,1]}]
  ]
```

A few examples may help you to understand all this. For a third-order Butterworth filter, we have

```
In:
Components[Butterworth[3]]

Out:
{1, 1., 2., 1., 1}
```

For the third-order Chebyshev filter with $\varepsilon = 0.4$, we have

```
In:
Components[Chebyshev[3,0.4]]

Out:
{1, 1.73284, 1.06569, 1.73284, 1}
```

Note that even-order Chebyshev filters do not have equal termination resistances:

```
In:
Components[Chebyshev[4,0.4]]

Out:
{1, 1.80704, 1.15193, 2.51308, 0.828301, 2.18163}
```

Although knowing the filter component values is useful, we are still a long way from building a filter. The component values currently returned by the **Components** function are for low-pass filters in a 1 Ω system with a cutoff frequency of $\omega = 1$. Fully realized filters need component values scaled to the frequency and system impedance at which the filter will operate; band-pass or high-pass filters require an additional transformation.

8.2 Transformations

We have to consider three transformation parameters that affect filter component values: impedance, frequency, and filter type. In Sections 8.2.1 through 8.2.4, we discuss how you can implement, with *Mathematica*, the transforms that are necessary to take account of these parameters.

8.2.1 Impedance Scaling

The impedance transformation must scale our $1\,\Omega$ system to any impedance level we want: Typically, system impedances of 50 or 75 Ω are chosen for high-pass and low-pass filters. We effect the impedance transformation by multiplying impedances by the system Z_0 and dividing admittances by Z_0. Our transform function must know whether the first element is an impedance, as shown in Figure 8.1, or an admittance (shunt capacitor), so that the multiplications and divisions, by Z_0, are performed correctly. Using a third-order filter as an example helps us demonstrate the transformation functions. Remember that **g[0]** and **g[4]** are the termination resistors. Historically, normalized filter elements have been denoted by **g**s — you should not mistake the **g**s for conductances.

```
In:
  comps = Map[g, Range[0,4]]
```
```
Out:
  {g[0], g[1], g[2], g[3], g[4]}
```

Case 1: If the first component is a series inductance, we want

```
In:
  comps {Zo, Zo, 1/Zo, Zo, Zo}
```
```
Out:
                        g[2]
  {Zo g[0], Zo g[1], ----, Zo g[3], Zo g[4]}
                        Zo
```

Case 2: If the first component is a shunt capacitance, we want

```
In:
  comps {Zo, 1/Zo, Zo, 1/Zo, Zo}
```
```
Out:
              g[1]          g[3]
  {Zo g[0], ----, Zo g[2], ----, Zo g[4]}
              Zo            Zo
```

By using a list of powers to invert or not to invert Z_0, we can obtain the result for each case:

```
In:
  len = Length[comps] - 2
```
```
Out:
  3
```

```
In:
  powers = Map[ If[OddQ[#],1,-1]&, Range[len]]
```
```
Out:
  {1, -1, 1}
```

Once we join powers for each termination, we have our first case:

```
In:
  comps (Zo^Join[{1},powers,{1}])
```
```
Out:
                            g[2]
  {Zo g[0], Zo g[1], ----, Zo g[3], Zo g[4]}
                            Zo
```

For our second case, we just invert the powers:

```
In:
  comps (Zo^Join[{1},-powers,{1}])
```
```
Out:
              g[1]          g[3]
  {Zo g[0], ----, Zo g[2], ----, Zo g[4]}
            Zo             Zo
```

We can use this technique to build a **Transform** function. We set the start parameter to **L** or **C** to indicate whether the first component is a series inductor or shunt capacitor. We then change the **L** or **C** value to a 1 or –1 by substitution. (Using **start_:L** tells *Mathematica* to expect an expression called **start**, but, if **Transform** is called with only two arguments, then start is to be assumed to have the value **L**.)

Get number of filter components - 2 terminations.
Set up alternating powers to multiply or divide by Zo.
Determine if first component is series or shunt.
Multiply out lists.

```
In:
  Transform[comps_List,Zo_,start_:L] :=
    Module[{len,powers,first},
       len = Length[comps] - 2;
       powers = Map[If[OddQ[#],1,-1]&,
                                  Range[len]];
       first = start /. {L->1,C->-1};
       comps (Zo^Join[{1},first powers,{1}])
    ]
```

So now **Transform** performs impedance scaling:

```
In:
    Transform[comps, Zo, L]
Out:
                          g[2]
    {Zo g[0], Zo g[1], ----, Zo g[3], Zo g[4]}
                          Zo
```

```
In:
    Transform[comps, Zo, C]
Out:
              g[1]            g[3]
    {Zo g[0], ----, Zo g[2], ----, Zo g[4]}
              Zo              Zo
```

8.2.2 Frequency Scaling

In comparison to impedance scaling, frequency scaling is easier. To implement frequency scaling, we divide all reactive components by the product of 2π and the filter's corner frequency. By modifying **Transform** to perform frequency scaling, we then have a complete low-pass filter solver in *Mathematica*:

```
In:
    Transform[comps_List,Zo_,start_:L,ws_] :=
        Module[{len,powers,first,wScale},
            len = Length[comps] - 2;
            powers = Map[If[OddQ[#],1,-1]&,
                                    Range[len]];
            first = start /. {L->1,C->-1};
            wScale = Join[{1},
                            Table[1/ws,{len}],
                            {1}];
            comps (Zo^Join[{1},
                            first powers,
                            {1}]) wScale
        ]
```

Set up list of radian frequencies to divide all components except terminations.

Multiply out lists.

By using *Mathematica*'s symbolic functionality, you can demonstrate fundamental aspects of filter design. For example, by calling **Transform** with symbolic arguments, you can show that the frequency scaling is applied only to the inductors and capacitors:

```
In:
  Transform[comps, Zo, L, 2 Pi fo]
```

```
Out:
          Zo g[1]      g[2]        Zo g[3]
{Zo g[0], -------, ----------, -------, Zo g[4]}
          2 fo Pi  2 fo Pi Zo   2 fo Pi
```

A basic 10 MHz low-pass filter has component values as shown:

```
In:
  b3 = Transform[Components[Butterworth[3]],
                 50, L, 2 Pi 10 MHz]
```

```
Out:
             -6        -9         -6
        2.5 10     2. 10     2.5 10
  {50, --------, -------, --------, 50}
          Pi        Pi        Pi
```

But the result is easier to read if we return a data type with labels. In the list **b3**, the nature of each component is not obvious — are the list's elements in farads or henries? Labels will make the list easier to read and allow us to format the list as a table. Also, the factor of **Pi** should be reduced to a number, with **N**, when you format the output. You should not use **N** in the **Transform** function, because you will limit the available precision. Each time you set up a data type you should add a definition to the **Normal** function to extract the data. This is a simple *but important* task because it keeps the *Mathematica* function set uniform in effect.

Overload Format to recognize a FilterForm data type.
Display the filter values as a table.

```
In:
  Format[FilterForm[x_]] :=
      EngineeringForm[
          N[TableForm[
              Prepend[x,{"Comp","Value"}]
          ]]
      ]

  Normal[FilterForm[y_List]] ^= y
```

```
Out:
  y
```

Now your results are easily read, and you can still extract the data for further use:

In:
FilterForm[Transpose[{{Rs,L,C,L,RL},b3}]]

Out:

Comp	Value
Rs	50.
L	795.775×10^{-9}
C	636.62×10^{-12}
L	795.775×10^{-9}
RL	50.

In:
Normal[%]

Out:

$$\{\{Rs, 50\}, \{L, \frac{2.5\ 10^{-6}}{Pi}\}, \{C, \frac{2.\ 10^{-9}}{Pi}\}, \{L, \frac{2.5\ 10^{-6}}{Pi}\},$$

$$\{RL, 50\}\}$$

We implement the labeling of component types using the **powers** variable
and a substitution:

In:
```
Transform[comps_List,Zo_,start_:L,ws_] :=
   Module[{len,powers,first,wScale,lbls},
      len = Length[comps] - 2;
      powers=Map[If[OddQ[#],1,-1]&, Range[len]];
      first = start /. {L->1,C->-1};
      wScale = Join[{1},Table[1/ws,{len}],{1}];
      lbls = Join[{Rs},
                      first powers /. {1->L,-1->C},
                      {RL}];
      FilterForm[
         Transpose[{lbls,
            comps(Zo^Join[{1},
                              first powers,
                  {1}]) wScale}]
      ]
   ]
```

Get labels by substitut-
ing L or C for the corre-
sponding power value.
Also add in Rs or RL for
the terminations.

Wrap the results and the
labels with the Filter-
Form name.

In:

```
Transform[Components[Butterworth[3]],
         50, L, 2 Pi 10 MHz]
```

Out:

Comp	Value
Rs	50.
L	$795.775 \ 10^{-9}$
C	$636.62 \ 10^{-12}$
L	$795.775 \ 10^{-9}$
RL	50.

8.2.3 High-Pass Transformation

Our **Transform** function needs the ability to create high-pass and band-pass filters from the low-pass form. This filter-type transformation works by transforming the frequency variable and ultimately results in changing the component type.

Imagine our brick-wall filter function in terms of *s*, the Laplace variable:

In:
```
H[s_] := If[ Abs[s] <= 1, 1, 0]
```

In the *s*-plane our filter passes frequencies within the unit circle, as shown by the white area in the following density plot:

In:
```
DensityPlot[ Re[
              H[sigma + I w]
            ],
            {sigma,-2,0},
            {w,-2,2},
            PlotLabel->"Low Pass H[s]",
            AxesLabel->{"sigma","w"},
            AspectRatio->Automatic,
            PlotPoints->30];
```

Out:

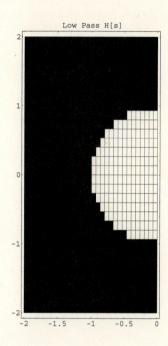

If we transform *s* to 1/*s*, all capacitors will become inductors and vice versa. Transforming *s* to 1/*s* maps the inside of the unit circle to the outside, as shown in the next plot:

In:

```
DensityPlot[ Re[
              H[1/(sigma + I w)]
           ],
           {sigma,-2,0},
           {w,-2,2},
           PlotLabel->"High Pass H[1/s]",
           AxesLabel->{"sigma","w"},
           AspectRatio->Automatic,
           PlotPoints->30];
```

Out:

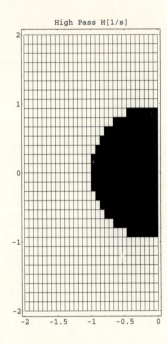

Our high-pass transformation changes an impedance of value sL to an impedance of value L/s. The impedance of L/s is realized as a capacitor of value $1/L$. For a high-pass filter **Transform** must invert each component value and change the labels on the components. At the end of Section 8.2.4 we shall incorporate the code for the high-pass transformation into a revised **Transform** function.

8.2.4 Band-Pass Transformation

The band-pass transformation is the last one that we shall consider. Of course, there are many other transformations, such as band-stop, which are discussed in books on filter design but, as we are mainly interested in showing the principles of applying *Mathematica* to filter design, we shall end our discussion of the subject with band-pass filters.

The band-pass transformation maps s to $(s + 1/s)$, so changing an inductor to a series resonator and a capacitor to a shunt resonator. By plotting $H(s + 1/s)$ with **DensityPlot**, we can see how band-pass areas are formed along the $I\omega$ axis at $\omega = \pm 1$:

In:
```
DensityPlot[H[(sigma + I w) + 1/(sigma + I w)]//Re,
           {sigma,-2,0},{w,-2,2},
           PlotLabel->"Band Pass H[s+1/s]",
           AxesLabel->{"sigma","w"},
           AspectRatio->Automatic,
           PlotPoints->30];
```

Out:

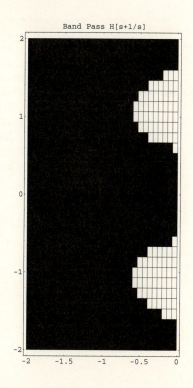

Actually, real band-pass transformations involve two scaling parameters: one for the center frequency and one for the Q, or selectivity (narrowness), of the filter. We shall use **LP**, **HP**, and **BP[Qo]** for our filter-type designators and include the Q in the band-pass designation as **Qo**. By changing how **Transform** works with components, we include high-pass and band-pass mappings in **Transform**. We start with the frequency-varying components from the input list of components. Low-pass filters need only scale filter component values by frequency. High-pass filters need to invert the component values and then scale each by frequency. Band-pass filters multiply the original components by Q and then combine both low-pass and high-pass realizations. The final impedance scaling will need to be a part of the band-pass transformation because it is different in the low-pass and high-pass cases.

In:

```
Transform[comps_List,Zo_,start_:L,ws_,type_]:=
    Module[{len,powers,first,lbls,rhp,rlp,
                    ele,result,r1,r2},
      len = Length[comps] - 2;
      powers = Map[If[OddQ[#],1,-1]&,
                            Range[len]];
      first = start /. {L->1,C->-1};
      ele = Take[comps,{2,len+1}];
          (* first & last *)
      {r1,r2} = comps[[{1,-1}]];
      If[Head[type] === BP,
          ele *= type[[1]];   (* extract Q *)
          rhp=(Zo^(-first powers))/(ele ws);
          rlp = (Zo^(first powers)) ele/ws;
          lbls = Join[{Rs}, first powers /.
                      {1->LC,-1->CL},{RL}]
        result = Join[{{Rs,Zo r1}},
              Transpose[{first powers /.
                      {1->LC,-1->CL},
                      rlp,rhp}],
                      {{RL,Zo r2}}];,
        If[type === HP,
          result = (Zo^(-first powers))/
                      (ele ws);
          result = Join[{Zo r1},
                      result, {Zo r2}];
          lbls = Join[{Rs},
                      first powers /.
                      {-1->Lhp,1->Chp},{RL}];
          result = Transpose[{lbls,
                          result}];,
          result = (Zo^(first powers))
                          ele/ws;
          result = Join[{Zo r1},
                      result, {Zo r2}];
          lbls = Join[{Rs},
                      first powers /.
                      {1->L,-1->C},{RL}];
          result=Transpose[{lbls,result}];
        ]
      ];
      FilterForm[result]
    ]
```

Get number of reactive comps.

Set up powers of 1 or -1 to multiply or divide by Zo.

First element is series (L) or shunt (C) and determines if powers list starts with 1 or -1.

Extract first and last comps as r1 and r2.

Multiply the circuit reactive elements (ele) by Qo.

Scale the HP elements (rhp) and the LP elements (rlp) by Zo and ws.

Use powers and a substitution to create the labels.

Join together the result matrix.

The HP and LP scaling only requires Zo and ws.

Wrap the result with FilterForm for formatting.

In:
Transform::usage=
"Transform[comps,Zo,startComp,wo,type]
returns a table of filter element values
as a FilterForm data type. Parameters include
a list of component values that begin and
end with source and load resistances, the
system impedance, Zo, the starting component
for the low-pass equivalent (L or C), the
radian frequency for scaling, and the
transformation type (LP, HP, or BP[Qo]).
The band-pass type should include a number
for Qo.";

The **usage** statement makes our transform function accessible to other people and provides the text that will form the reply to a request for on-line help.

To see **Transform** in action, calculate a high-pass, third-order Butterworth filter:

In:
Transform[Components[Butterworth[3]],
        50, L, 2 Pi 10 MHz, HP]

Out:

Comp	Value
Rs	50.
Chp	$318.31 \ 10^{-12}$
Lhp	$397.887 \ 10^{-9}$
Chp	$318.31 \ 10^{-12}$
RL	50

Because *Mathematica* has symbolic abilities, you can remind yourself of the transformation formulae by using **Transform** with symbols:

```
In:
Transform[ {1,g1,g2,g3,1}, zo, L, wo, BP[Qo]]

Out:
   Comp    Value

   Rs      zo

           g1 Qo zo              1
           --------      -----------
   LC          wo        g1 Qo wo zo

           g2 Qo            zo
           -----         --------
   CL      wo zo         g2 Qo wo

           g3 Qo zo              1
           --------      -----------
   LC          wo        g3 Qo wo zo

   RL      zo
```

The ideal band-pass transformation is used rarely because the transformation gives unwieldy component values for very narrow-band filters. Most of the art of filter design is concerned with how to realize a filter in the desired frequency range: Top-coupled, helical, and crystal band-pass filters are commonly used for very narrow bandwidths [Zverev67, Rhodes76].

8.3 Basic Synthesis

Filtering is one of the few areas in electronic engineering in which we can really design a circuit. Most engineering design is design by analysis: We take specifications, assume a circuit topology based on experience, and, through repeated analysis, obtain a circuit that works. Filter design consists of two parts: approximation and synthesis. The approximations of Butterworth and Chebyshev polynomials to a brick-wall transfer function represent two solutions to the approximation problem. In synthesis, we take our approximated transfer function and derive component values analytically. (We skipped over the synthesis process in Section 8.1 because we used formulae for our element values.) Often we have only our transfer function approximation and need to derive component values from that approximation.

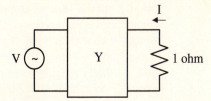

Figure 8.2   Singly terminated network. I is current in this diagram.

8.3.1 Singly Terminated Synthesis

Almost all syntheses require a network input impedance, rather than a transfer function, from which we can derive element values. A first step is to find the simplest network that has its transfer function directly related to an input impedance. Singly terminated networks have transfer functions in terms of transfer impedances or admittances that we can easily translate to input or output impedances [Kuo66]. A singly terminated network is designed to have a load resistor but can be driven by a voltage or current source: Operational amplifiers or operational transconductance amplifiers work nicely for driving these networks. Figure 8.2 shows a singly terminated network defining a transfer admittance, Y_{21}. Y_{21} is defined in terms of Y-parameters (note that I refers to current, rather than $\sqrt{-1}$ in this equation):

$$Y_{21} = \frac{I}{V} = \frac{y_{21}}{1 + y_{22}} = \frac{P(s)}{M(s) + N(s)} \quad .$$

If the network is lossless, our synthesis procedure is straightforward. $M(s)$ will be an even function of s, whereas $N(s)$ will be an odd function of s. $P(s)$ may be even or odd in terms of s. Each of our Y-parameters must be a ratio of even to odd or odd to even polynomials for the lossless condition to hold. To find a solution, we need to divide through by $M(s)$ or $N(s)$ so that y_{21} satisfies the lossless condition — we shall then find y_{22}. We can then expand y_{22} directly into component values by using a continued fraction expansion. For example, if $P(s)$ was even then y_{21} would be $P(s)/N(s)$ and y_{22} would be $M(s)/N(s)$. Our third-order Butterworth filter would have the transfer function $H(s) = 1/(1 + 2\,s + 2\,s^2 + s^3)$ and $y_{22} = (1 + 2\,s^2)/(2\,s + s^3)$.

8.3.2 Component Values by Continued Fraction Expansion

A continued fraction expansion converts a ratio of two polynomials into a sequence of terms involving s and remainders. The idea is similar to the

ContinuedFraction function example in Stephen Wolfram's book [Wolfram91], but this time we use polynomials. Each remainder is inverted, and another term is removed until there is no remainder left. In terms of circuit elements, the continued fraction expansion amounts to removing components to form a ladder network. The components create transfer-function zeros at infinity. First, we shall manually perform, with *Mathematica*'s help, a continued fraction expansion on a third-order Butterworth filter's y_{22}, and then we shall construct an algorithm:

$$y_{22} = \frac{1 + 2s^2}{2s + s^3}$$

To remove terms of s, we must invert y_{22} and then use **PolynomialQuotient**:

```
In:
  PolynomialQuotient[(2 s + s^3),(1 + 2 s^2),s]
```
```
Out:
  s
  -
  2
```

The answer is a term in s and so represents a series inductance. The coefficient of s gives the value for the inductor nearest the load, since we are using y_{22}. We also need to find the remainder:

```
In:
  PolynomialRemainder[(2 s + s^3),(1 + 2 s^2),s]
```
```
Out:
  3 s
  ---
  2
```

Now, we invert again and divide this remainder into the original numerator to give a shunt capacitance:

```
In:
  PolynomialQuotient[(1 + 2 s^2), 3 s/2, s]
```
```
Out:
  4 s
  ---
  3
```

```
In:
    PolynomialRemainder[(1 + 2 s^2), 3 s/2,s]
Out:
    1
```

Lastly, we invert again and factor out a series inductor:

```
In:
    PolynomialQuotient[3 s/2, 1, s]
Out:
    3 s
    ---
     2
```

```
In:
    PolynomialRemainder[3 s/2, 1,s]
Out:
    0
```

When the remainder is zero, we are done; we list the components in **vals**. Note that the last result is now first, in keeping with the left-to-right ordering of components in the low-pass Butterworth filter shown in Figure 8.3.

```
In:
    vals = {3 s/2, 4 s/3, s/2};
```

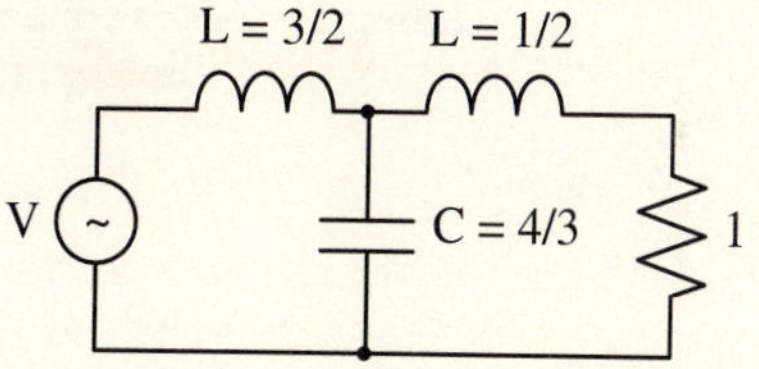

Figure 8.3 Singly terminated Butterworth filter.

The continued fraction expansion can be reconstructed from the components by **FoldList**. **FoldList** is used with a pure function that inverts the last result and adds the new result to that value. The last term in the list returned by **FoldList** is the continued fraction expansion of $1/y_{22}$.

```
In:
    FoldList[(#2 + 1/#1)&, Infinity, vals]
Out:
                  3 s    2     4 s   s        1
    {Infinity,    ---,   --- + ---,  - + ---------}
                   2     3 s    3    2    2     4 s
                                         --- + ---
                                         3 s    3
```

```
In:
   Simplify[1/Last[%]]
```

```
Out:
            2
   1 + 2 s
   --------
            3
   2 s + s
```

We can use the **Nest** function to create our continued fraction expansion automatically. To use **Nest**, we need a function that returns the polynomial quotient and remainder in a form that can be nested. **cfeProcess** will return the quotient, the numerator and denominator polynomials, and the variable, so that repeated applications yield a continued fraction expansion. Note that **cfeProcess** works on a list.

```
In:
   cfeProcess[{result_,num_,denom_,var_}] :=
       {PolynomialQuotient[num, denom, var], denom,
        PolynomialRemainder[num, denom, var], var}
```

Now **NestList** can be used three times, corresponding to the maximum power of *s*. The first element of each list gives our components:

```
In:
   NestList[cfeProcess,
           {dummy,(2 s + s^3),(1 + 2 s^2),s}, 3]
```

```
Out:
                         3              2       s          2   3 s
   {{dummy, 2 s + s , 1 + 2 s , s}, {-, 1 + 2 s , ---, s},
                                                2          2

       4 s  3 s            3 s
   {---, ---, 1, s}, {---, 1, 0, s}}
     3    2              2
```

We can define a function to perform the entire component calculation. **Exponent** is used to derive the highest power (assumed to be in the numerator) of *s* in y_{22}; the largest exponent determines how many times **NestList** must be used. Using **Rest**, we drop the first list from **NestList** — the first result was simply the starting value and is not needed. We then extract the first element of each result list because the first elements contain the component values. Finally, we remove insignificant numbers (that creep in when computers manipulate real numbers) from the list by using **Chop**. Note that the component values proceed from the output to the input, since this is y_{22}, the *output* admittance.

Get the highest power of
var, n.
Do the CFE n times.

Extract just the component.

```
In:
  CFELadder[num_,denom_,var_] :=
    Module[{n,tmp},
            (* highest power *)
        n = Exponent[num,var];
        tmp = NestList[cfeProcess,
                            {dummy,num,denom,var},
                        n];
        Chop[Expand[First /@ Rest[tmp]] ]
    ]

In:
  CFELadder[(2 s + s^3),(1 + 2 s^2),s]

Out:
    s   4 s   3 s
  {-,  ---,  ---}
    2    3     2
```

The continued fraction expansion allows us to convert transfer impedances and admittances into a ladder network. The preceding **CFELadder** function is useful only for low-pass networks, but we can easily extend it to other kinds of networks. It is fascinating how such a simple mathematical operation can convert an abstract transfer function into concrete inductor and capacitor values for a network!

The synthesis discussed in this section on basic synthesis is limited in two ways: It is limited to ladder networks and to singly terminated networks. Most commercially produced filters are designed to work in systems of 50 or 75 Ω impedance. (Requiring that a filter be driven by a voltage or current source is often inconvenient and nearly impossible at very high frequencies.) Also, doubly terminated filters match the prototype networks for many digital filters. We discuss doubly terminated networks and the more general synthesis technique of pole-zero extraction in Section 8.4.

8.4 Advanced Synthesis

Much of practical filter design involves tables of component values and specific realization techniques that are highly optimized for loss, Q, bandwidth, size, weight, or frequency range. Most designers who use filters have a favorite set of circuits for various problems. Designers often avoid custom filters because they incorrectly regard synthesizing arbitrary filters as a task for experts. We will explore synthesis by both insertion loss and optimization methods.

Actually, the principles of filter synthesis, as currently practiced, were developed in 1939 by Darlington [Darlington39], when advanced computer-aided tools were not available. The lack of inexpensive software to aid engineers in filter synthesis is one of the main reasons that filters have remained in the realm of experts. Thankfully, *Mathematica* has the root-finding routines, library of advanced math functions, and programmability that are needed to make synthesis accessible to everyone.

8.4.1 Doubly Terminated Synthesis

Darlington established a comprehensive theory for synthesizing lossless networks, using the insertion loss of a network, the typical starting point for a designer, to derive an input impedance that was synthesizable. The link between the input impedance and the insertion loss is formed by the S-parameters. For a lossless network, the equation $S_{11}(s)\,S_{11}(-s) + S_{12}(s)\,S_{12}(-s) = 1$ holds, where S_{12} $(= S_{21})$ is our transfer function and $S_{11}(s) = (Z_{in}(s) - Z_0)/(Z_{in}(s) + Z_0)$. The preceding equation relating S_{11} and S_{12} was derived in Chapter 4. By using S-parameters, we ensure that our system is doubly terminated. Now let $S_{21}(s) = H(s) = P(s)/Q(s)$ and then we can solve for $S_{11}(s) = M(s)/N(s)$:

$$S_{11}(s)\,S_{11}(-s) \;=\; \frac{Q(s)\,Q(-s) - P(s)\,P(-s)}{Q(s)\,Q(-s)}$$

The hard part of finding $S_{11}(s)$ is computing the roots of the numerator and denominator of $S_{11}(s)\,S_{11}(-s)$. Once we have found the roots, we can collect the left-half plane roots to form $M(s)/N(s)$, and a little algebra gives us an equation for Z_{in}. Once we have Z_{in}, a continued fraction expansion gives the component values:

$$Z_{in}(s) \;=\; \frac{1 + S_{11}(s)}{1 - S_{11}(s)}$$

$$=\; \frac{N(s) + M(s)}{N(s) - M(s)}$$

8.4.2 Spectral Factorization

Spectral factorization is the process for finding the roots of $S_{11}(s)\,S_{11}(-s)$. We can find the roots just as we did in Section 8.1, although now we shall be more general:

```
In:
  p2 = Numerator[ h2B[3,s] ]
```

```
Out:
  1
```

```
In:
   q2 = Denominator[ h2B[3,s] ]
```

```
Out:
         6
   1 - s
```

```
In:
   rN = Solve[q2 == 0, s]
```

```
Out:
                              1/3                 2/3
   {{s -> 1}, {s -> (-1)    }, {s -> (-1)    }, {s -> -1},

                 4/3                 5/3
      {s -> (-1)    }, {s -> (-1)    }}
```

```
In:
   lhpN = Select[ Map[ (s /. #)&, rN],
                        (Re[N[#]] < 0)&]
```

```
Out:
            2/3           4/3
   {(-1)    , -1, (-1)    }
```

Constructing *N*(*s*) is easy, once we have the left-half plane roots:

```
In:
   n = Apply[Times, s + lhpN]
```

```
Out:
                    2/3           4/3
   (-1 + s) ((-1)    + s) ((-1)    + s)
```

The roots of *M*(*s*) are tricky to extract because often they will have zero real parts. Roots with real parts of zero exist in pairs and are not selected by our test function for **lhpN**. Next we will develop a way of selecting half the roots on the *I*ω axis:

```
In:
   m2 = q2 - p2
```

```
Out:
         6
   -s
```

```
In:
   rM = Solve[m2 == 0, s]
```

```
Out:
   {{s -> 0}, {s -> 0}, {s -> 0}, {s -> 0},
    {s -> 0}, {s -> 0}}
```

Since the roots of $M(s)\,M(-s)$ along the $I\omega$ axis will not always be at zero, we must carefully extract one-half of the roots and associate that half with $M(s)$. The other half of the roots is associated with $M(-s)$. By selecting those roots with zero real parts, we can count the unique roots and keep half of them. Actually, our function will find the unique roots, count the number of each root, and then reconstruct a list of roots using one-half of the number counted:

axis contains those roots near the $I\omega$ axis.

```
In:
  axis = Select[ Map[ (s /. #)&, rM],
                         (Abs[Re[N[#]]] < 10^-9)&]
Out:
  {0, 0, 0, 0, 0, 0}
```

Union reduces the axis list to unique numbers.

```
In:
  u = Union[axis]
Out:
  {0}
```

```
In:
  c = Map[ Count[axis,#]&, u ]
Out:
  {6}
```

Now the unique root values found by **Union** are duplicated according to one-half of their count, by applying **Table**:

```
In:
  Apply[Table[#1,{#2}]&, Transpose[{u,c/2}],1]
Out:
  {{0, 0, 0}}
```

Flatten will convert this result to a simple list of roots. The use of **Union**, **Count**, and **Table** may seem overly complicated in this case, yet it will prove essential for synthesizing filters other than Butterworth. The **HtoZin** function below constructs $Z_{in}(s)$ directly from $H(s)\,H(-s)$. *Mathematica* has enabled us to use very little code to solve quite a large problem.

Many papers on synthesizing filters by computer are devoted to numerical precision and mappings to enhance precision. We have ignored the precision issue, because **Solve** works so well and our filters are small. The one obvious place in the **HtoZin** function where precision may be a problem is where roots with real part magnitudes less than 10^{-9} are associated with the $I\omega$ axis:

```
In:
HtoZin[s21sq_, var_] :=
   Module[{rM,rN,q2,p2,m2,lhp,n,m,num,u,c},
          p2 = Numerator[ s21sq ];
          q2 = Denominator[ s21sq ];
          rN = Solve[q2 == 0, var];
          lhp = Select[Map[(var /. #)&,rN],
                           (Re[N[#]] < 0)&];
          n = N[Apply[Times, s - lhp] ];
          m2 = q2 - p2;
          rM = Solve[m2 == 0, var];
          lhp = Select[Map[(var /. #)&,rM],
                           (Re[N[#]] < 0)&];
          axis = Select[Map[(s /. #)&,rM],
                           (Abs[Re[N[#]]] < 10^-9)&];
          u = Union[axis];
          c = Count[axis,#]& /@ u;
                (* duplicates roots from Count/2 *)
          axis = Flatten[
                      Apply[ Table[#1,{#2}]&,
                            Transpose[{u,c/2}],1]];
          m = N[ Apply[Times, s-Join[lhp,axis]] ];
          Chop[Expand[(n+m)]]/Chop[Expand[(n-m)]]

          ]
```

Extract the left-half plane roots of q2 into lhp. Construct the n polynomial with the lhp roots and s as the variable.

Find the roots for m2. Extract the left-half plane roots. Then extract all the roots on the $I\omega$ axis.

Use Union, Count, and Table to put the LHP roots and half the $I\omega$ axis roots into axis.

We use **Chop** and **Expand** to get the polynomials in the right form and to remove insignificant numbers.

8.4.3 A Butterworth Example

Now, we can construct an input impedance for a doubly terminated Butterworth filter:

```
In:
zB = HtoZin[ h2B[3,s], s ]

Out:
                   2       3
1. + 2. s + 2. s  + 2 s
-------------------------
                   2
1. + 2. s + 2. s
```

Once we have the input impedance, we can perform a continued fraction expansion to get the component values:

```
In:
   CFELadder[ Numerator[zB],
                Denominator[zB], s]
```

```
Out:
   {1. s, 2. s, 1. + 1. s}
```

Here, we have the components ordered from the input to the output. The output component also includes the termination resistor of $1\,\Omega$. Note that these component values are not the same as those resulting from singly terminated synthesis. There is no mistake here, even though each synthesis used a third-order Butterworth filter: The doubly terminated filter uses two termination resistors, which influence the pole locations and cause a change in the component values.

8.4.4 Pole-Zero Extraction

All networks are not ladders. Also, all poles will not be at DC or infinity. So a more general approach takes the input impedance and extracts elements in a given sequence [Saal58]. The extraction sequence depends on pole-zero information in the transfer function, as well as the input impedance. A complete discussion of synthesis would include consideration of realizability conditions, Brune sections, zero shifting, and more. Yet our work with continued fraction expansions implicitly uses pole extraction to synthesize a network. By exploring pole-zero extraction a bit more, we can make the idea of pole-zero removal more concrete and prepare ourselves to use *Mathematica* in more advanced filter-synthesis problems.

Going back to our first synthesis example, we had a y_{22} for a low-pass filter:

$$y_{22} = \frac{1 + 2s^2}{2s + s^3} = \frac{P(s)}{Q(s)}$$

To extract a first pole at infinity, we want to remove a term in s. So we invert y_{22} and remove an impedance, because s^3 is the dominant term as s goes to infinity. Using **Series** to expand our polynomial at **Infinity** gives us the pole directly:

```
In:
   Series[(2 s + s^3)/(1 + 2 s^2),{s,Infinity,2}]
```

```
Out:
   s      1 0
   - + O[-]
   2      s
```

The residue from the pole at infinity gives a coefficient, k, of 1/2 and the reduced polynomial ratio:

```
In:
  new = Together[(2 s + s^3)/(1 + 2 s^2) - s/2]
```

```
Out:
      3 s
   ------------
             2
   2 (1 + 2 s )
```

The reduced polynomial ratio is just our beginning polynomial ratio minus the pole we found using **Series**. Our pole of $s/2$ is an inductor of 1/2 henries, and our reduced polynomial ratio is the leftover input impedance. This inversion and pole removal at infinity will give us the same component values as the continued fraction expansion.

If we have a high-pass network, our pole removal will be at DC rather than infinity and the extracted component values will be proportional to $1/s$ to remove poles at DC. The following transfer impedance has three zeros at DC, so network pole removal should happen at DC [Kuo66]:

$$Z_{21} = \frac{z_{21}}{1 + z_{22}} = \frac{s^3}{s^3 + 3s^2 + 4s + 2} \quad ,$$

$$z_{22} = \frac{s^3 + 4s}{3s^2 + 2}$$

To remove a $1/s$, we invert z_{22} and use **Series** at 0. The first term represents a 2 H shunt inductor:

```
In:
  Series[(1 + 2 s^2)/(2 s + s^3), {s,0,2}]
```

```
Out:
    1     3 s         3
   --- + --- + O[s]
   2 s    4
```

Removing the pole at zero gives us a reduced polynomial. By continuing the inversion and pole-removal process, we derive a high-pass network:

```
In:
  newHP = Together[(1 + 2 s^2)/(2 s + s^3)-1/(2 s)]
```

```
Out:
      3 s
   ----------
            2
   2 (2 + s )
```

Many networks contain resonators that have poles or zeros at finite frequencies. Poles and zeros at finite frequencies have to be removed carefully, or negative element values will occur during synthesis [Temes77]. You can use the method of partial pole removal to prevent negative element values from occurring during synthesis. The transfer impedance given next is used to show partial pole removal. Note that Y_{21} defines a low-pass network with a zero at $s = \sqrt{-1/2}$.

$$Y_{21} = \frac{y_{21}}{1 + y_{22}} = \frac{2 + 4s^2}{2 + 2s + 6s^2 + 5s^3} \quad ,$$

$$y_{22} = \frac{2 + 6s^2}{2s + 5s^3}$$

We need to remove a transmission zero at infinity, and a transmission zero at the finite frequency of $s = \sqrt{-1/2}$. If we start by removing a pole at infinity, by inverting y_{22} to get an impedance, we get into trouble:

```
In:
  Series[(2 s + 5 s^3)/(2 + 6 s^2), {s,Infinity,2}]

Out:
  5 s       1 0
  --- + O[-]
   6         s
```

The residue from the pole at infinity gives a coefficient, k, of 5/6, and the leftover input impedance is given by

```
In:
  newBP =
      Together[(2 s + 5 s^3)/(2 + 6 s^2) - 5 s/6]

Out:
          s
  -------------
              2
  6 (1 + 3 s )
```

which we can invert to give an admittance, which is a simple shunt resonator:

```
In:
  Simplify[ 1/newBP ]

Out:
  6
  - + 18 s
  s
```

This appears to be a shunt network that gives us a zero at DC that we do not have and also does not satisfy our need for a transmission zero at $s = \sqrt{-1/2}$. What we really want to do is to remove just enough inductance so that the leftover impedance has a zero at $\sqrt{-1/2}$. If we set the frequency to $\sqrt{-1/2}$, our input impedance must be just the desired inductor:

```
In:
  X1 = (2 s + 5 s^3)/(2 + 6 s^2) /. s->Sqrt[-1/2]
```

```
Out:
   5 I
  ---- - I Sqrt[2]
   3/2
  2
```

Since **X1** is equal to sL_1 at $s = \sqrt{-1/2}$, L_1 is easy to find:

```
In:
  L1 == Simplify[ X1/Sqrt[-1/2] ]
```

```
Out:
          1
  L1 == - 
          2
```

Now, we can remove our first inductor. Since L_1 is less than 5/6 — our value for completely removing the pole at infinity — extracting L_1 is a partial pole removal:

```
In:
  newZ = Together[(2 s + 5 s^3)/(2 + 6 s^2) - s/2]
```

```
Out:
            3
    s + 2 s
  -----------
             2
  2 (1 + 3 s )
```

Next, we remove our transmission zero, a shunt-resonant admittance, and finally we can remove the rest of the pole at infinity. The series-resonant network has the admittance **lastY**, which we must remove from 1/**newZ** so the admittance pole (transmission zero) at $\sqrt{-1/2}$ is removed completely. We know that $L_2 C_2 = 2$, so we simplify our analysis by assigning $L_2 C_2 = 2$. The LC resonator admittance is given below as **yRes**. Then we use **yRes** to get **lastY** by letting **L2 -> 2/C2** — leaving only **C2** as a variable:

In:
```
yRes = s C2/(s^2 L2 C2 + 1);

lastY = Together[1/newZ - (yRes /. L2->2/C2)]
```

Out:
```
         2        2
 2 + 6 s    - C2 s
 ----------------
         2
   s (1 + 2 s )
```

To cancel the pole at $\sqrt{-1/2}$ we need $6 - C_2 = 4$, or $C_2 = 2$. That makes $L_2 = 1$. Substituting in all these values leaves us with the last component, another inductor of 1/2 H:

In:
```
Simplify[lastY /. C2->2]
```

Out:
```
  2
  -
  s
```

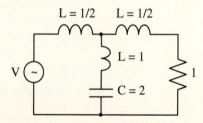

Figure 8.4 Synthesized filter with $I\omega$ axis zero.

Figure 8.4 shows our final filter. Direct synthesis is not always easy. Often, a filter needs to be built quickly without a lot of research. The next section takes you through a practical filter design using optimization and *Nodal*.

8.4.5 Design by Optimization

Not all design can be done by synthesis. As CAD programs have become readily available and computers have become more powerful, more engineers are using optimization as a design method. In some cases design by optimization is an excuse to avoid thinking. In other cases it is a well thought out design procedure [Orchard85, Carlin77]. In the following design example we use *Nodal* and *Math-*

ematica to design a custom filter by a combination of analytic design and optimization.

The filter specification in Figure 8.5 can be approximated by the following set of data points, **spec**. The shaded regions of the filter specification set limits for the transfer function. The specification shows that at least 20 dB of rejection is needed below 50 MHz. The 50 MHz point needs at least 40 dB of rejection. The passband of the filter begins at 75 MHz. Less than 1 dB of attenuation is required to 100 MHz with up to 3 dB attenuation above 110 MHz. The data points are in MHz and dB.

```
In:
    spec = {{25,-20},{50,-40},{75,0},
                            {85,0},{100,-1}};

    {frs,gains} = Transpose[spec];
```

Filter specification

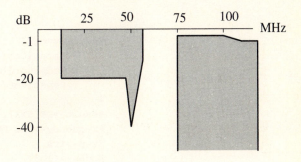

Figure 8.5   High-pass filter specification.

We will use analytic formulae to estimate the initial filter values. Since the filter characteristic is a high-pass, we can start our design with values for a 65 MHz Butterworth high-pass filter. We shall use the **Components** and **Transform** functions from previous sections to compute the Butterworth values. We start with a third-order filter:

```
In:
    Transform[
            Components[Butterworth[3]],
            50, L, 2 Pi 65 MHz, HP]
```

```
Out:
   Comp   Value

   Rs     50.

                    -12
   Chp    48.9708 10

                    -9
   Lhp    61.2134 10

                    -12
   Chp    48.9708 10

   RL     50.
```

You can analyze the Butterworth high pass with the following *Nodal* code. Obviously, there is not enough rejection at 50 MHz and a little too much rejection at 75 MHz.

```
In:
Needs["Nodal`","Nodal2.m"];

NodalAnalyze[
        NodalNetwork[
                Capacitor[{1,2}, 49 pF],
                Inductor[{2,0}, 61 nH],
                Capacitor[{2,3}, 49 pF]
        ],
        Result->DB[S21],
        Frequency->frs MHz,
        Step->frs,
        Nodes->{1,3}
]
```

```
Out:
   Step   DB[S21]

   25.    -24.9

   50.    -7.69

   75.    -1.55

   85.    -0.805

   100.   -0.323
```

Next we need to do two things: make sure our filter attenuates by 40 dB at 50 MHz and optimize the response. To make sure we get 40 dB attenuation, we can put a capacitor in series with the shunt inductor: The series LC circuit can resonate at 50 MHz and force at least 40 dB of attenuation. Next we must optimize the rest of the response into our specification.

To make optimization work best, you should minimize the number of variables. You can make the input and output capacitors equal because the network is symmetric. You can also compute the capacitor value in the shunt arm from the inductor value. The capacitor computation also forces the resonance to stay at 50 MHz. You can use the following **CVal** function to compute the resonant capacitance from an inductor value:

In:
```
CVal[L2_] := N[1/((2 Pi 50 MHz)^2 L2)]
```

You should be careful about how you compute the error between your circuit response and your specifications. If you leave all the numbers in dB then an error of 1 dB in the stopband is as important as an error of 1 dB in the passband — yet 1 dB in the passband is far more serious. You should use the gain magnitude rather than the gain dB to compute your error. Using magnitudes will give a better weighting to the passband and stopband performance. The next equation converts the dB specification to magnitudes for the error function:

In:
```
mags = ArcDB[gains]
```
Out:
```
{0.1, 0.01, 1., 1., 0.891251}
```

In:
```
tmp = NodalAnalyze[
        NodalNetwork[
            Capacitor[{1,2}, 49 pF],
            Inductor[{2,4}, 61 nH],
            Capacitor[{4,0}, CVal[61 nH] ],
            Capacitor[{2,3}, 49 pF]
        ],
        Result->Abs[S21],
        Frequency->frs MHz,
        Step->frs,
        Nodes->{1,3}
    ]
```

```
Out:
  Step    Abs[S21]

  25.     0.106

                    -14
  50.     1.14 10

  75.     0.471

  85.     0.675

  100.    0.862
```

You can compute the error at each frequency point by subtracting the *Nodal* results from the specification, as shown below. The maximum error occured at 75 MHz.

```
In:
  mags - Transpose[Normal[tmp]][[2]]
```

```
Out:
  {-0.0064546, 0.01, 0.529165, 0.325487, 0.0288879}
```

The following error function puts together several pieces of the preceding analysis. It executes *Nodal* with the given capacitor, **c1**, and inductor, **12**, values and subtracts the results from the design specifications in **mags**. Finally, the total squared error is computed by a dot product of the error with itself. The error vector, **err**, contains an error value at each frequency point.

```
In:
    error[c1_,12_,frs_,mags_] :=
      Module[ {tmp,err},
          tmp=NodalAnalyze[
              NodalNetwork[
                      Capacitor[{1,2}, c1 pF],
                      Inductor[{2,4}, 12 nH],
                      Capacitor[{4,0}, CVal[12 nH] ],
                      Capacitor[{2,3}, c1 pF]
                  ],
              Result->Abs[S21],
              Frequency->frs MHz,
              Step->frs,
              Nodes->{1,3}
          ];
          err = mags - Transpose[Normal[tmp]][[2]];
          err.err
      ]
```

You should check the beginning error value produced by the starting component values. If the starting error is too low your optimization will terminate too easily. As shown here our starting error is acceptable at 0.39:

```
In:
  error[ 49, 61, frs, mags]

Out:
  0.386934
```

Finally, you can use **FindMinimum** to solve for component values. If more than a few components are used, **FindMinimum** can take a long time to converge. **FindMinimum** reduces the error to 0.0015, which appears to be very good. You should also check that the component values are reasonable — and the components are reasonable in this case:

```
In:
  FindMinimum[error[c1,l2,frs,mags],
              {c1,10,100},{l2,30,250}]

Out:
  {0.00150981, {c1 -> 28.8596, l2 -> 184.128}}
```

Now that you have optimized component values and a low error, you can plot the final result and compare it with the specification:

29 pF   29 pF
(1) ┤├ (2) ┤├ (3)
        184 nH
(4)
        CVal[184 nH]
(0)

Use Epilog to plot the specification points with the desired point size.

```
In:
  NodalPlot[NodalAnalyze[
              NodalNetwork[
                  Capacitor[{1,2}, 29 pF],
                  Inductor[{2,4}, 184 nH],
                  Capacitor[{4,0},CVal[184 nH]],
                  Capacitor[{2,3}, 29 pF]
              ],
              Result->DB[S21],
              Frequency->Range[5,120] MHz,
              Step->Range[5,120],
              Nodes->{1,3}
          ],
          PlotRange->{-45,0},
          Epilog->Prepend[Point /@ spec,
                          PointSize[0.025]]
  ]
```

Out:

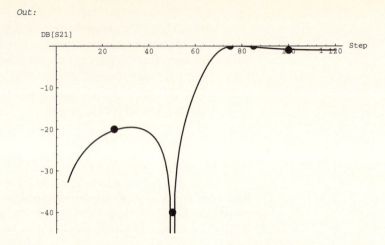

The plot shows the final filter response, with points at the error specifications. Even though few specification points were used, the filter appears to meet the specifications of Figure 8.5 at all the frequencies. By using few specification points, minimizing the number of variables, and choosing good starting values and topology for your design, you made optimization an efficient design technique.

You should make two changes to the preceding code for production filter design. First, you should make the error function return zeros when the response is within the tolerance window. The present code tries to match the specifications exactly — which is a simpler, but less useful, criterion. Second, you should set your optimization specifications somewhat tighter than the actual requirements so that component value variations do not move filters out of specification.

That wraps up our tour of analog filter design. Although engineers often use analog-filter prototype values to design digital filters directly, many digital filters have unique design procedures. *Mathematica* is well suited both to implementing digital-filter design procedures and to working with digital filters and sampled signals.

8.5 Digital Filtering

Analog filters rely on precise and unchanging component values to meet their design specifications consistently. High-precision, stable components are often expensive and are difficult to keep stable during temperature fluctuations and as they age. The digital filter, which can be either implemented in hard-wired logic or programmed into a microprocessor, offers a solution to these problems by periodically digitizing (sampling) the analog signal and then numerically processing the samples. The phenomenon of sampling (and introductory digital filters) is well explained by Lynn [Lynn82].

In this section, we introduce you to using *Mathematica* for designing digital filters. We demonstrate aliasing, mapping from the *s*- to the *z*-plane, converting from analog-filter specifications to digital-filter coefficients, and designing finite impulse response digital filters.

8.5.1 Sampling of Signals

You can use two of *Mathematica*'s simplest plotting functions to demonstrate the effect of sampling on an analog signal. The sampled form requires you to evaluate the signal at instants in time and results in a list of values, one for each sample taken. You can create this list using **Table**. Note how the signal appears to be part of a slowly varying sine wave with a period roughly twice the length of the section plotted. Using a sine wave as the continuous analog signal, you can plot its continuous (analog) form with **Plot** and then show the continuous and sampled forms together using **Show**:

In:
```
sample = ListPlot[Table[Sin[6 t],
                  {t,1,10}],
                  AxesOrigin->{0,-1},
                  AxesLabel->{"sample","value"},
                  Prolog->PointSize[0.025]]
```

Out:

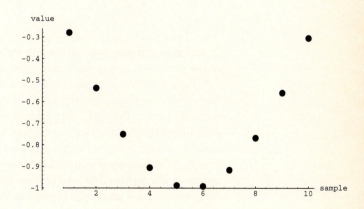

In:
```
function = Plot[Sin[6 t],{t,1,10}]
```

Out:

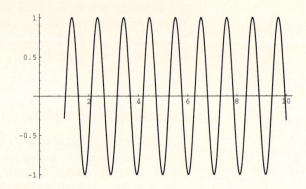

In:
```
Show[sample, function]
```

Out:

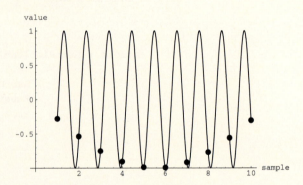

In the combined plot, you can see how the samples pick successively earlier parts of the sine wave, giving the wrong impression of the latter's frequency.

For a fixed sampling frequency, sine-wave signals with frequency up to one-half of the sampling frequency will be sampled correctly. As the signal frequency increases further, the signal appears to be decreasing in frequency until, when the sampling and signal frequencies are equal, it appears to be a DC signal. At frequencies higher than the sampling frequency, the cycle of illusion repeats: Signals with frequencies in commensuration with the sampling frequency will appear as DC signals. The s-plane cannot be used for representing sampled signals, since there is an infinite number of indistinguishable locations along the $j\omega$-axis that appear to represent the same signal. (With a sampling rate of ω_s, signals at $2\omega_s$, $3\omega_s$, . . . all appear as DC ($\omega = 0$) signals, for example.)

8.5.2 Mapping of *s*- to *z*-Plane

The *z*-plane is a suitable alternative to the *s*-plane because it represents a mapping from the Cartesian *s*-plane to a *polar* coordinate system: the cyclic nature of the sampled signal's apparent frequency is mapped onto the (cyclic) angle coordinate.

Positions in the *s*-plane (representing signals with different σ and ω components, as described in Chapter 7) map onto the *z*-plane such that the *s*-plane's complex axis becomes a circle of unit radius on the *z*-plane: the most easterly point on the circle representing the position of signals with frequencies that are an integer multiple of the sampling frequency (including DC). Proceeding counterclockwise around this unit circle, points on the unit circle represent signals of increasing frequency until the most westerly point ($\omega = \omega_s/2$), corresponding to the Nyquist frequency, is reached. Continuing around the unit circle, signals appear to have a decreasing frequency (although their real frequency would be increasing) until, at the most easterly point where, apparently, only DC signals would be present. Signals with $\sigma < 0$ are represented inside the unit circle, and those with $\sigma > 0$ are represented outside.

You can define an *s*- to *z*-plane mapping function **szMap** that takes two arguments (the σ and ω components of an *s*-plane position) and converts them to an $\{r, \theta^\circ\}$ position in the *z*-plane. You can map this function over a list of complex numbers to see where they map:

```
In:
  szMap[sw_Complex] :=
        ComplexToPolarDegree[N[Exp[sw]]];
  Map[szMap,{I,2I,3I,4I,5I,6I,7I}]//N

Out:
  {{1., 57.2958}, {1., 114.592}, {1., 171.887},
  {1., -130.817}, {1., -73.5211}, {1., -16.2253},
  {1., 41.0705}}
```

8.5.3 Infinite Impulse Response Filters

Digital filters process signals by adding together a selection of sample values after they have each been multiplied by some coefficient. Just as you need to calculate component values for analog filters, so do digital filters require the calculation of these coefficients. In earlier sections of this chapter, you saw how to generate filter transfer functions in the *s*-plane. You can use *Mathematica* to help you convert the analog-filter transfer functions into the digital filter's coefficients, which are used by digital filters; the translation of analog-filter designs to their digital-filter equivalents is a large subject. In this section, we show you how an infinite impulse response filter's coefficients can be calculated using one of the many techniques available (see [Terrell80] for further details). (Finite impulse response filters are normally designed using the techniques that we shall introduce in Section 8.5.4.)

The bilinear transform is one of the principal techniques for carrying out the transform from the analog s-domain transfer function to the digital z-domain equivalent. For example, if you wish to convert an analog second-order Butterworth low-pass filter to its digital equivalent then you need to make two substitutions: first, replacement of s by s/ω_c (for a low-pass filter, with a sampling interval of t, ω_c is the cutoff frequency in the digital filter and is given by $\omega_c = 2/t \tan(\omega(t/2))$, and second, by $(2/t)(z - 1)/(z + 1)$. Other filter types have different replacement forms. A second-order Butterworth filter has poles at $s = (-\sqrt{2}, j\sqrt{2})$ and $(-\sqrt{2}, -j\sqrt{2})$ and has a transfer function $H(s) = 1/(s^2 + s\sqrt{2} + 1)$. This s-domain transfer function will become $H(z)$ — named **hz** in our example. By supplying numerical values for t (the sampling interval, from a 600 Hz sampling frequency) and f (the frequency of the low-pass filter, 100 Hz in this example), you can determine the equation for the delay-domain transfer function:

```
In:
  hs = 1/(s^2 + Sqrt[2] s + 1);
  % /. s->s/wc;
  hz = % /. s->(2/t)((z-1)/(z+1))
```

```
Out:
                      1
      -----------------------------------
                    2        3/2
            4 (-1 + z)      2    (-1 + z)
      1 + --------------- + -------------
              2   2             2
             t  wc  (1 + z)    t wc (1 + z)

             t   wc  (1 + z)
```

```
In:
  rule1 = wc->(2/t)Tan[Pi f t];
  rule2 = s->(2/t)((z-1)/(z+1));
  zz = ((hz /. rule1) /. rule2);
  Together[ExpandAll[% /. {f->100, t->1/600}]] //N
```

```
Out:
                      2
            (1. + z)
      ---------------------------
                               2
      1.55051 - 4. z + 6.44949 z
```

There are many other techniques for designing digital filters. One simple technique is to place poles and zeros manually in the *z*-plane and to iterate on their precise position by comparing the resulting frequency response with your filter specification. Manual pole-zero positioning is not the most efficient way to design a digital filter, but it does give you a good feeling for the effects that different pole and zero positions and combinations have on a filter's transfer function. *Mathematica* can help you with manual filter design by removing the tedium of calculation and letting you concentrate on visualizing the results of your designs.

The magnitude of a digital filter's transfer function is the product of the lengths of the vectors from each zero to the point on the unit circle (which represents the frequency at which the filter is being evaluated) divided by the product of the lengths of the vectors from each pole. The arithmetic complexity of this ratio can be handled for you by *Mathematica*. The following example takes two lists — one of the *z*-plane zeros and one of the *z*-plane poles — and plots the amplitude of the filter's transfer function. Each pole or zero is specified by its polar coordinates, $\{r, \theta^{\text{rads}}\}$ where $r = 1$ for a position on the unit circle and (for a stable filter) $r < 1$ for positions inside the unit circle. θ^{rads} is zero at DC and π at one-half of the sampling frequency f_s; $\theta^{\text{rads}} = 2\pi f/f_s$. Next, the transfer function's numerator (**tfN**) and denominator (**tfD**) are calculated. Once **tfN** and **tfD** are known, **tf** (the ratio of the products of the vectors from each zero and each pole to the point on the unit circle representing the frequency at which the transfer function is being evaluated) is calculated:

```
In:
  zeroList = {{1,Pi},{-1,Pi}};

  poleList = {{0.95,0.5},{0.95,-0.5}};

  tfN = Map[(z-(#[[1]] Cos[#[[2]]]+
            #[[1]] Sin[#[[2]]] I))&, zeroList];

  tfD = Map[(z-(#[[1]] Cos[#[[2]]]+
            #[[1]] Sin[#[[2]]] I))&, poleList];

  tf[zIN_List]:=N[Product[Abs[tfN[[i]]],
                          {i,Length[tfN]}]/
                Product[Abs[tfD[[i]]],
                        {i,Length[tfD]}]
              /. z->(zIN[[1]] Cos[zIN[[2]]]+
                 I zIN[[1]] Sin[zIN[[2]]])];

  Plot[ tf[{1,theta}], {theta,0,Pi},
            AxesLabel->{"2πf/fs","tf"}]
```

Out:

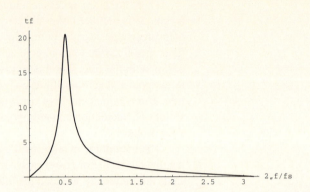

To calculate the z-domain transfer function, you calculate the symbolic ratio of **tfN/tfD**. You can then determine the time-domain form of the transfer function by equating a term in z^{-n} with a sample taken n time steps ago. If the filter is to be workable, the transfer function should contain z to only negative powers (since the filter does not have access to samples not yet taken). You can achieve this by simply dividing numerator and denominator by z^m, where m is the highest positive power:

In:
```
Together[ExpandAll[Product[tfN[[i]],
                           {i,Length[tfN]}]/
                   Product[tfD[[i]],
                           {i,Length[tfD]}]]];
Chop[%]
```

Out:
$$\frac{-1 + z^2}{0.9025 - 1.66741\,z + z^2}$$

You can implement this filter using *Mathematica*, too. First, you need to create a signal to filter in an array **randomData** and then create a null-valued array for the result in **op**:

In:
```
randomData = Table[Random[],{1024}];
Short[randomData]
```

Out:
```
{0.598226, 0.421905, <<1021>>, 0.960539}
```

```
In:
  op = Table[0,{Length[randomData]}];
  For[ i=1,i<=Length[randomData],i++,
      If[i>2,
          op[[i]] = -0.9025 op[[i-2]]+
            1.66741 op[[i-1]]+
            randomData[[i]] -
            randomData[[i-2]],
          op[[i]] = 0]];
  Short[op]
```

```
Out:
  {0, 0, -0.33806, <<1019>>, <<7>>17, -1.03487}
```

As an alternative to the procedural technique with the **For** loop, here is a functional approach. The function **ar** is set up for **FoldList**, which provides the feedback, or memory, that is required for autoregressive work; **ar** creates a list containing the new output and the last output value. **FoldList** repeatedly applies **ar** to the last computation and the next input value. The function **Partition** is used to make a matrix in which each row contains a number and a number that is two increments away, so the moving average can work directly on this with **Map**:

```
In:
  ar[0,x_] := {x,0}
  ar[xL_List,x_] :=
  {1.66741 xL[[1]] - 0.9025 xL[[2]] + x, xL[[1]]}

  arma[data_] :=
  Module[{tmp},
          tmp = Map[(#[[3]]-#[[1]])&,
                    Partition[data,3,1]];
          First /@ Drop[ FoldList[ar,0,tmp] ,1]
          ]

  op2 = arma[randomData];
  Short[op2]
```

```
Out:
  {-0.33806, -0.820212, <<1019>>, -1.03487}
```

If your computer supports *Mathematica*'s sound facilities, you can hear how the filter affects the original data by listening to **randomData** and **op**:

```
In:
  ListPlay[randomData, SampleRate->2048,
          PlayRange->All]
```

Out:

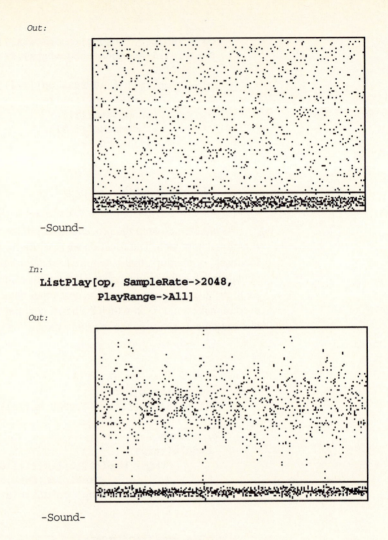

-Sound-

In:
```
ListPlay[op, SampleRate->2048,
        PlayRange->All]
```

Out:

-Sound-

8.5.4 Finite Impulse Response Filters

Since multiplication in the Fourier domain is equivalent to convolution in the time domain, a filter function that is defined in the frequency domain and then Fourier transformed to calculate the time-domain equivalent will produce the time series with which a signal's time series can be convolved to perform filtering.

You can use *Mathematica*'s Fourier transform function to generate simple finite impulse response filters. The first step is to specify the transmission of the

filter as a function of frequency, remembering that the specification must be symmetrical about the Nyquist frequency.

The following list represents a filter that has unity transmission from DC to
one-quarter of the sampling frequency:

```
In:
    filterFunction =
        {1,1,1,1,0.5,0,0,0,0,0,0,0,0.5,1,1,1}

Out:
    {1, 1, 1, 1, 0.5, 0, 0, 0, 0, 0, 0, 0, 0.5, 1, 1, 1}
```

To obtain the time-domain series, we transform **filterFunction** using
the **Fourier** command; *Nodal*'s **FourierEE** command automatically
invokes **Chop**:

```
In:
    Fourier[filterFunction]
```

N

```
Out:
                                     -18
    {2., 1.25683 - 1.72117 10     I, 0.,
                              -18
      -0.374151 + 2.09387 10     I, 0.,
                             -19
       0.167045 - 8.74138 10     I, 0.,
                             -18
      -0.0497281 + 3.7405 10     I, 0.,
                              -19
      -0.0497281 + 6.09864 10     I, 0.,
                             -19
       0.167045 + 3.18484 10     I, 0.,
                             -18
       0.374151 + 1.98545 10     I, 0.,
                             -18
       1.25683 - 6.15285 10     I}
```

```
In:
    filterFT = FourierEE[filterFunction]

Out:
    {1., 0.628417, 0, -0.187076, 0, 0.0835223, 0, -0.024864}
```

For a filter of a prescribed length (**len**), it is necessary to apply a windowing
function that attenuates high frequencies due to the truncated filter that would
otherwise result in undesirable sidelobes appearing in the filter's transmission
function. There are many functions that can be used as window functions. In the
next example, we choose a Hamming window for use with a nine-term filter:

```
In:
  HammingWindow[i_,len_] :=
                0.54 + 0.46 Cos[Pi i/len]
  windowCoefs=
          Table[N[HammingWindow[i,4]],{i,0,4}]
```

```
Out:
  {1., 0.865269, 0.54, 0.214731, 0.08}
```

Because both the filter and the window coefficients have now been computed, we can calculate the actual filter coefficients from their product:

```
In:
  filterCoefs = Table[
          windowCoefs[[j]] filterFT[[j]],
          {j,1,Min[Length[filterFT],
                  Length[windowCoefs]]}]
```

```
Out:
  {1., 0.54375, 0, -0.0401709, 0}
```

This list contains one-half of the finite impulse response filter: from the midpoint (of value 1.0) to the right end. To obtain symmetrical coefficients, we need to reverse the last four coefficients and then to place them in front of those obtained so far.

Toolbox

Manipulating Lists

Reverse[*list*] reverses the order of the elements in *list*.

Drop[*list*, *n*] returns *list* with the first *n* elements removed. If *n* is negative, then the last *n* elements are removed.

Join[*list1*, *list2*,...] concatenates its arguments into one list.

```
In:
  Reverse[filterCoefs]
```

```
Out:
  {0, -0.0401709, 0, 0.54375, 1.}
```

However, the midpoint should not be repeated — so the **1.** is removed from the end of the filter before the two lists are joined:

In:
  Drop[%,-1]

Out:
  {0, -0.0401709, 0, 0.54375}

In:
  Join[%,filterCoefs]

Out:
  {0, -0.0401709, 0, 0.54375, 1.,
      0.54375, 0, -0.0401709, 0}

8.6 Summary

In this chapter, you have seen how

* to use many internal *Mathematica* functions and write custom functions for creating filters
* to compute element values for many filter types
* to manually extract elements from a ratio of polynomials or pole-zero removal
* to create filters from a specification
* to map from the *s*-plane to the *z*-plane
* to convert analog filter designs to their digital counterparts
* to design finite impulse response filters from a specified spectrum.

8.7 Exercises

8.1 Create an *s*-plane three-dimensional plot of a Butterworth filter. Use surface color to represent phase information. The surface height represents magnitude information. (*Hint*: Check **Options[Plot3D]**.)

8.2 Use **FindRoot** to solve for the minimum filter order for a given attenuation. Compare the Butterworth and Chebyshev solutions.

8.3 Design three band-pass filters with a 50 MHz center frequency in a 50 Ω system, and Qs of 1, 10, and 100. Comment on the element values and the effect of possible element parasitics on the filter performance.

8.4 Make **Transform** use options for **Zo** and **Type**. Leave **freq** as an argument. This new **Transform** will be simpler to use since it has fewer items that must be specified.

8.5 Write a function that finds the nearest ±5% tolerance commercial component to the filter value computed; ±5% values use the list {1, 1.1, 1.2, 1.3, 1.5, 1.6, 1.8, 2, 2.2, 2.4, 2.7, 3, 3.3, 3.6, 3.9, 4.3, 4.7, 5.1, 5.6, 6.2, 6.8, 7.5, 8.2, 9.1}.

8.6 Include a band-stop filter type in **Transform**.

8.7 Include Bessel filters in **Components** (*see* [Kuo66]).

8.8 Include elliptic filters in **Components**. How must **Transform** change?

8.9 Use **CFELadder** on the transfer impedance $Z_{21}(s) = 2/(s^3 + 3s^2 + 4s + 2)$.

8.10 Use **CFELadder** on the transfer impedance $Z_{21}(s) = s^3/(s^3 + 3s^2 + 4s + 2)$.

8.11 Modify **CFELadder** to extract transfer impedances or admittances.

8.12 Modify **CFELadder** to operate on a low-pass transfer function, rather than the separated numerator and denominator of y_{22}.

8.13 Modify **CFELadder** to extract high-pass networks (poles at DC).

8.14 Use **CFELadder** and **HtoZin** to derive a doubly terminated Chebyshev filter with a ripple factor of 0.4.

8.15 Use a different optimization method to derive a set of final component values for the filter in Section 8.4.5 (*see* [Temes77] or [Orchard85]).

8.16 Modify the transfer-function magnitude calculation for arbitrary pole-zero location (in Section 8.5.3) to include the relative phase shift for the output signal.

8.8 References

[Baher84] Baher, H., *Synthesis of Electrical Networks*, John Wiley & Sons, Chichester, United Kingdom, 1984.

[Carlin77] Carlin, H. J., "A New Approach to Gain-Bandwidth Problems," *IEEE Trans. CAS*, pp. 170–175, April 1977.

[Darlington39] Darlington, S., "Synthesis of Reactance-Four-Poles which Produce Prescribed Insertion Loss Characteristics," *J. Math. Phys.*, 30:257–353, 1939.

[Jacobs78] Jacobs, G.M., Allstot, D.J., Broderson, R.W., and Gray, P.R., "Design Techniques for MOS Switched Capacitor Ladder Filters," *IEEE Trans CAS*, 12:1014–1021, 1978.

[Kuo66] Kuo, F.F., *Network Analysis and Synthesis*, John Wiley & Sons, New York, 1966.

[Lynn82] Lynn, P.A., *An Introduction to the Analysis and Processing of Signals*, Macmillan, London, 1982.

[Matthaei80] Matthaei, G. L., Young, L., and Jones, E. M. T., *Microwave Filters, Impedance-Matching Networks, and Coupling Structures*, Artech House, Dedham, MA, 1980.

[Orchard85] Orchard, H. J., "Filter Design by Iterated Analysis," *IEEE Trans. MTT*, pp. 1089–1096, November 1985.

[Rhodes76] Rhodes, J.D., *Theory of Electrical Filters*, John Wiley & Sons, London, 1976.

[Saal58] Saal, R., and Ulbrich, E., "On the Design of Filters by Synthesis," *IRE Trans. Circuit Theory*, CT-5, December 1958.

[Temes77] Temes, G.C., and LaPatra, J.W., *Introduction to Circuit Synthesis and Design*, McGraw-Hill, New York, 1977.

[Terrell80] T. J. Terrell, *Introduction to Digital Filters*, Macmillan, New York, 1980.

[Wolfram91] Wolfram, S., *Mathematica – A System for Doing Mathematics by Computer*, Addison-Wesley, Redwood City, California, 1991.

[Zverev67] A. I. Zverev, *Handbook of Filter Synthesis*, John Wiley & Sons, New York, 1967.

CHAPTER 9

High-Frequency Circuits and Analysis

Computer-aided techniques are especially valuable when measurement techniques can change circuit response. When you measure a circuit with an oscilloscope probe, a CAD program can predict the effects of the loading so that you separate the real response from the measurement. Practical high-frequency circuit analysis is more difficult than low-frequency circuit analysis because the circuit components cannot be treated as lumped. At high frequencies, component sizes (for example, conducting paths on printed circuit boards) approach the wavelength of the frequencies handled by the circuit and it becomes necessary to consider intercomponent connections as transmission lines with end reflections which you have to model and minimize. In this chapter, we show how *Mathematica* can help you tackle some of these problems.

9.1 The Smith Chart

The Smith chart is primarily a tool for converting impedances into reflection coefficients. Typically, the reflection coefficients are based on voltages, but, as we shall see, current-based reflection coefficients are also useful. Phillip Smith invented this chart back in 1939 [Smith39], and his leap of insight provided a fundamental tool for generations of RF engineers. Today, the chart is used by engineers for matching impedances, for displaying transistor characteristics, and for optimizing amplifier gain, power, and noise figure specifications.

9.1.1 Impedance and Reflection

To best understand the Smith chart, we start with how you can use it to calculate the reflection coefficient that results from a signal propagating between two impedances. Given an impedance, $z = r + \mathrm{I}\,x$, and a system impedance, z_0, we can convert them to a voltage-reflection coefficient, ρ, by the formula, $\rho = (z - z_0)/(z + z_0)$. We can implement the reflection coefficient formula — along with its inverse — using *Mathematica*:

```
In:
  rho[z_,zo_:50] := (z - zo)/(z + zo)

  zFromRho[rho_, zo_:50] := zo (1 + rho)/(1 - rho)
```

In **zFromRho**, we assign a default system impedance, **zo**, of 50 Ω by the syntax **zo_:50**. The formula **rho** defines a bilinear transformation that maps straight lines in the impedance plane into circles in the reflection plane.

You can visualize this bilinear transformation by defining lines of constant resistance and reactance in the impedance plane and mapping them onto the reflection plane. Our first step is to define the lines of constant resistance and reactance as coordinate pairs so that we can plot them using *Mathematica*'s **Line** graphics function.

You need to know about several of *Mathematica*'s graphics primitives in order to create and use Smith charts.

Toolbox

Graphics Functions

Line, **Point**, **PointSize**, **Circle**, and **Text** are some of the graphics primitives native to *Mathematica*. You can create lines from a list of points (for example, **Line[{pt1, pt2,...}]**, where each point is a pair of coordinates such as **{x1,y1}**).

The **Point** primitive draws a point at coordinates **{x1,y1}**:
Point[{x1,y1}].

The current point size is set using **PointSize[*val1*]**, where *val1* is a fraction of the plot width.

Circle[{*x1,y1*},*r*] creates a circle of radius *r* at {*x1,y1*}. You can also use **Circle[{*x1,y1*}, *r*, {*theta1*, *theta2*}]** to draw an arc of radius *r* from angle *theta1* to angle *theta2*. The angle arguments are in radians, and *theta2* must be greater than *theta1*.

Text["*label1***", {*x1,y1*}]** centers the string "*label1*" at {*x1,y1*}

Primitives create a set of specifications used by **Show**.

Line, **Point**, **PointSize**, **Circle**, and **Text** work as wrappers with **Graphics** and **Show**, in the same way that **Resistor**, **Capacitor** work with **NodalNetwork** and **NodalAnalyze**.

Both **NodalNetwork** and **Graphics** work as wrappers containing components (or primitives). Both **NodalAnalyze** and **Show** operate on the objects and primitives.

Each **Graphics** object contains a list of primitives. **Show** can operate on a single **Graphics** object or take many **Graphics** objects as arguments.

ArcTan

You can use **ArcTan** with one or two arguments. When you use **ArcTan**[*x1*,*y1*], there is no ambiguity in the angle, as there is when you call **ArcTan** with one argument.

Note that we apply various scaling functions to the resistance and reactance coordinates so that the curves will be evenly spaced in the reflection domain. The constant-resistance lines are vertical and the constant-reactance lines are horizontal in the impedance plane:

```
In:
    rLines = Table[ 50 {10^r, Sinh[x]},
                {r,-1.5,1,0.5}, {x,-3,3,0.1}] //N;

    xLines = Table[ 50 {10^r, x},
                {x,-10,10,1}, {r,-1.5,1,0.25}] //N;
```

You can use **Show** to combine and plot the lines of constant resistance and reactance:

```
In:
    zChart = Show[ Graphics[{Line /@ rLines}],
                Graphics[{Line /@ xLines}]];
```

```
Out:
```

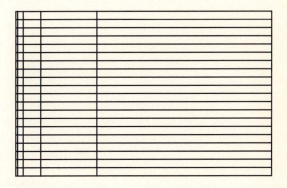

9.1.2 Generation of a Smith Chart

To map the lines of constant resistance and reactance onto the reflection plane, you need a function that will convert the real and imaginary parts of an impedance to the real and imaginary parts of a reflection coefficient. The function **rhoPairs** converts pairs of coordinates from impedance to reflection plane:

```
In:
  rhoPairs[{r_,x_},zo_:50] :=
      {Re[rho[r + I x,zo]],Im[rho[r + I x,zo]]}
```

You can then map the impedance points through a bilinear transformation: The first use of **Map** in **rhoR** gets the real and imaginary part of the transformed impedance at each point in a line; the second use of **Map** maps the transformation over each constant resistance line, thus creating the circles assigned to **circs**:

```
In:
  rhoR = Map[Map[ rhoPairs, #]&, rLines] //N;
  circs = Show[ Graphics[{Line /@ rhoR}],
                AspectRatio->Automatic]
Out:
```

The same mapping over constant-reactance lines produces arcs:

```
In:
  rhoX = Map[Map[ rhoPairs, #]&, xLines] //N;
  arcs = Show[ Graphics[{Line /@ rhoX}],
                AspectRatio->Automatic];
```

Out:

Once the reflection-plane circles and arcs are combined, you get a good representation of a Smith chart:

In:
```
rhoChart = Show[circs,arcs];
```

Out:

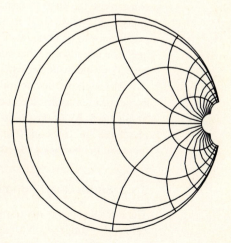

You can plot the impedance and Smith charts side by side using **GraphicsArray**:

In:

    Show[GraphicsArray[{zChart,rhoChart}]];

Out:

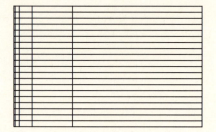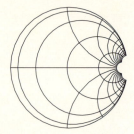

The center of the Smith chart represents zero reflection. A load impedance that is the conjugate of the source impedance creates no signal reflection and allows maximum power transfer (as we showed in Section 4.6.1). Typically, Smith charts are used with systems of 50 Ω source impedance — so a load impedance of 50 Ω would lie at the center of the chart and give zero reflection. You should be aware that many Smith charts are printed with an impedance of 1 Ω at their center and are described as normalized or per-unit.

The outer circle of the Smith chart corresponds to the left edge of the impedance rectangle. The other edges of the impedance rectangle map in order to the small cusps at the right side of the Smith chart.

For a given system impedance, you can plot any other impedance value on the chart and know what reflection caused it and vice versa. More importantly, you can use the chart to visualize the result of a series or parallel combination of impedances, as we shall do in Section 9.4. For now, you can imagine how adding a resistance to an impedance moves the impedance along a constant reactance line. You can also see that adding a reactance to an impedance moves the impedance along a constant resistance line.

Although our discussion gives insight into how impedances are transformed or mapped into reflection coordinates, it does not describe a good way to draw a Smith chart. All of the points and short line segments used to draw **zChart** and **rhoChart** are time consuming to draw. With *Mathematica*, the best way of drawing a Smith chart uses **Circle** and **Line**.

9.1.3 Smith Chart Function

The circles and arcs on a Smith chart represent constant-resistance or -reactance values [Gonzalez84]. Normalized (assuming the system impedance is 1 Ω) constant-resistance circles have a center at $r/(1 + r)$ and a radius of $1/(1 + r)$; normalized constant-reactance circles have a center at $\{1, 1/x\}$ and a radius of $1/x$.

The production of quality Smith charts requires high-resolution printers because the charts have hundreds of fine lines to enable engineers to accurately read values from the chart. Our Smith chart will have low resolution because we shall not have to read numbers from it (*Mathematica* will provide accurate numeric answers), and so we shall only need a few constant-resistance and constant-reactance lines. You can start with three normalized resistance and reactance values — {0, 1/3, 1, 3}.

Arcs should end on the outer circle of the Smith chart, which corresponds to a constant resistance of zero. You can use **Solve** to find all the arc intersections with the zero-resistance outer circle. All of the constant-reactance arcs have one point at the {1,0} coordinate in the reflection plane. Remember that the radius and center of our constant-reactance circle correspond with $1/x$ which equals 3 if x is 1/3. For the constant-reactance arc of value 1/3, the arc intersections are at {–4/5, 3/5} and {1,0} (where **u** and **v** in the following function correspond to the real and imaginary parts of a reflection coefficient):

The first equation is for the outer circle. The second equation is for the constant-reactance arc.

```
In:
   arcSoln1 = Solve[ {u^2 + v^2 == 1,
                      (u-1)^2 + (v-3)^2 == 3^2},
                      {u,v}]

Out:
                4             3
    {{u -> -(-), v -> -}, {u -> 1, v -> 0}}
                5             5
```

The **Circle** function needs the reflection coordinates converted to angles of arc. The {1,0} coordinate will be $\pi/2$ if the center is above {1,0}, or $-\pi/2$ if the center is below {1,0}. You can use *Mathematica*'s **ArcTan** function to find the arc angle between the arc center and the intersection coordinates. For the constant-reactance arc of value 1/3, you can calculate the following arc angles:

```
In:
   ang1 =
   N[ Apply[ArcTan,({u,v}-{1,3} /. arcSoln1),1]]

Out:
   {-2.2143, -1.5708}
```

Now you can show simultaneously a constant resistance circle of zero and a constant reactance arc of value 1/3:

```
In:
   Show[ Graphics[{Circle[{0,0}, 1],
                   Circle[{1,3}, 3, ang1]}],
         AspectRatio->Automatic];
```

Out:

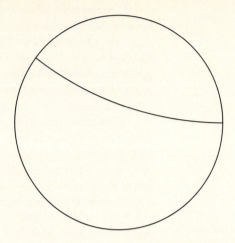

Once you solve for the other arcs in the series {1/3,1,3}, you can build a useful Smith chart function. You should also add a line through the middle of the chart, corresponding to the zero-reactance arc of infinite radius:

In:

```
smithG = Graphics[{
            Circle[{0,0}, 1],
            Circle[{0.25,0}, 0.75],
            Circle[{0.5,0}, 0.5],
            Circle[{0.75,0}, 0.25],
            Circle[{1,3}, 3, {-2.2143,-1.5708}],
            Circle[{1,-3}, 3, {1.5708,2.2143}],
            Circle[{1,1}, 1, {-3.14159,-1.5708}],
            Circle[{1,-1}, 1, {1.5708,3.14159}],
            Circle[{1,0.333},0.333,{-4.069,-1.5708}],
            Circle[{1,-0.333},0.333,{1.5708,4.069}],
            Line[{{-1,0},{1,0}}]}];
```

Next you will create a function for drawing a Smith chart and data (**SmithCharter**). Our **SmithCharter** function will draw a Smith chart and as many lists of points as you wish. We shall use **SmithCharter** again in Section 9.4 to trace component points along the Smith chart.

The **SmithCharter** function uses three underscores between **rhos** and **List** so that zero or more lists of points can be plotted on a Smith chart. The underscore is an important part of *Mathematica* syntax. For example, **myData_List** prescribes the variable **myData** to be a list of one or more values; **myData__List**, with two underscores, prescribes **myData** to be a sequence of one or more lists; **myData___List**, with three underscores, prescribes **myData** to be a sequence of zero or more lists. So, **rhos___List** declares an argument for **SmithCharter**, which can reference any number of lists, including the null case.

Options, if any are given, are passed through to **Show** so that you can use the standard graphics options. Usually you should filter the options so only those options recognized by **Show** are passed; the subject of filtering options is discussed by Maeder in *Programming in Mathematica* [Maeder90].

In:

```
SmithCharter::usage =
"SmithCharter[rhoLists, options] shows a
Smith Chart and plots the given points on
the chart.";

SmithCharter[ rhos___List, options___Rule ] :=
    Module[ {tmp},
        Show[ smithG,
            Graphics[ Join[ {PointSize[0.02]},
                Map[ Point[{Re[#],Im[#]}]&,
                    Flatten[{rhos}] ] ]
            ], Sequence @@ Prepend[{options},
            AspectRatio->Automatic
        ]
    ]
```

Show both the Smith chart graphic and a graphic of all the points given in rhos. Use Flatten to create one long list of points. Use Re and Im to get the real and imaginary coordinates of each point. Use Map to map Point over all the points.

You can plot a couple of random points with **SmithCharter** to see your function at work:

In:

```
SmithCharter[{-0.25 + I 0.5,-0.75 + I 0.5},
    PlotLabel->"Points on a Smith Chart"];
```

Out:

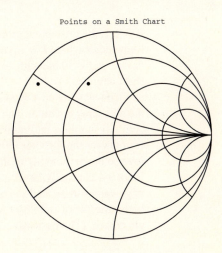

Points on a Smith Chart

The **SmithChart** function in *Nodal* contains many options and is a more complete function than the **SmithCharter** function. In Section 9.4, we shall work more with impedances and Smith charts.

9.2 Stability Analysis Using S-Parameters

When you use active devices, such as transistors, certain combinations of input and output loads can cause the device to oscillate. The device forward and reverse gains combine with load and source impedances to create a circuit with infinite gain at or near the frequency of oscillation. You can find the borders of the oscillation region by solving for those two-port input or output impedances that cause the reflection coefficients to have a magnitude of unity. These regions of the Smith chart can be mapped for every device operating frequency. The unstable regions shown on the Smith chart tell oscillator designers what imped-ances they have to achieve and amplifier designers what impedances they must avoid. It is very important for all designers to analyze device stability outside of the designed operating frequency range so that spurious oscillations do not occur.

The first task in stability analysis is to compute the Rollett stability factor, K [Gonzalez84]. K is given by:

$$K = \frac{1 + |\Delta|^2 - |s_{11}|^2 - |s_{22}|^2}{2|s_{12}s_{21}|} \quad,$$

where

$$\Delta = s_{11}s_{22} - s_{21}s_{12} \quad.$$

If $K \geq 1$, the unstable regions lie outside the Smith chart. For those devices and frequencies where $K \geq 1$, you have a well-defined maximum available gain and a device that is stable with any passive load a user may present, including open or short circuits.

You can create a *Mathematica* function for the K factor, **StabilityK**, from the preceding equation. The function **StabilityK** takes an S-parameter matrix of complex numbers and returns the K factor for a device at a given frequency. The formula for the K factor tells you a great deal about a stable device. In K, the inverse proportionality to the product $s_{21}s_{12}$ shows we need $s_{21}s_{12}$ to be small. Since s_{21} is usually large, a stable device usually has a small-valued s_{12}. Also, large values of s_{11} and s_{22} tend to make devices unstable. The K factor is not the best indicator of stability for a large network, but it is a good indicator for a sin-gle device. (The best way to analyze network stability is through the poles of a network. The *K* factor makes a simple evaluation of two-port stability that is usually correct.)

```
In:
  StabilityK[{{s11_,s12_},{s21_,s22_}}] :=
    Module[ {delta},
        delta = s11 s22 - s21 s12;
        (1+Abs[delta]^2 - Abs[s11]^2 - Abs[s22]^2)/
                        (2 Abs[s21 s12])
    ]
```

9.2.1 Device *K* Factor

We can illustrate the *K* factor, as well as the gain circles in Section 9.2.2, by using magnitude and angle S-parameter data from an NEC NE202 hetero-junction field effect transistor (FET). Our data range from 2 to 26 GHz, in 4 GHz steps:

```
In:
  fs = {2, 6, 10, 14, 18, 22, 26};
  s11m = {0.99,0.92,0.85,0.78,0.73,0.69,0.62};
  s11a = {-19,-53,-83,-110,-129,-140,-160};
  s21m = {3.47,3.11,2.79,2.52,2.24,2.01,1.77};
  s21a = {164, 133, 107, 82, 61, 46, 33};
  s12m = {0.03,0.06,0.09,0.11,0.11,0.1,0.1};
  s12a = {77, 61, 44, 25, 17, 16, 15};
  s22m = {0.67,0.63,0.58,0.56,0.53,0.53,0.51};
  s22a = {-11,-30,-50,-70,-80,-87,-107};
```

To use our **KFactor** function we must transform the magnitude and angle data from the device data book to matrices of complex numbers:

```
In:
  s11c = s11m E^(I s11a Degree) //N;
  s12c = s12m E^(I s12a Degree) //N;
  s21c = s21m E^(I s21a Degree) //N;
  s22c = s22m E^(I s22a Degree) //N;
```

Transpose is very useful for manipulating data. A matrix of lists becomes a list of matrices by transposing between levels in our data:

```
In:
  sMatrices =
          Transpose[{{s11c,s12c},
                     {s21c,s22c}},
                     {2,3,1}];
```

Mapping our **StabilityK** function over the data shows how *K* varies over our chosen range of frequencies:

In:
```
ListPlot[Transpose[{fs,
         Map[ StabilityK, sMatrices]}],
         PlotJoined->True,
         AxesLabel->{"GHz","K"}];
```

Out:

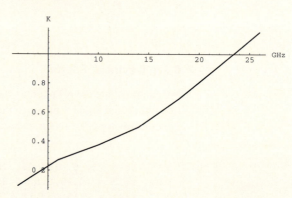

The device is conditionally stable over the region of the band where *K* is less than unity. Conditional stability does not make the device unusable: Devices with *K* factors less than unity are often made stable by a small amount of loss in the matching network. Typically, amplifiers are designed as a compromise between gain, noise figure, and output power. *K* factors less than unity indicate that the maximum available device gain is infinite.

In the next section we will use *Nodal* to examine the stability factor of various networks.

9.2.2 CAD and the Stability Factor

Most high-frequency CAD programs compute the stability factor, *K*, directly. *Nodal* can compute the stability factor for passive and active networks. Passive networks with low resistive losses have stability factors at or near unity no matter what their attenuation.

In:
```
Needs["Nodal`",Nodal2.m];

NodalAnalyze[NodalNetwork[
             Inductor[{1,2}, 10 nH],
             Capacitor[{2,0}, 2 pF]],
          Result->{DB[S21],K},
          Frequency->{1,5} GHz, Step->{1,5}]
```

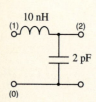

N

Out:
```
    Step    DB[S21]   K

    1.      -0.985    1.

    5.      -20.      1.
```

Lossy passive networks have stability factors greater than or equal to unity:

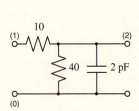

In:
```
NodalAnalyze[
        NodalNetwork[
                Resistor[{1,2}, 10],
                Resistor[{2,0}, 40],
                Capacitor[{2,0}, 2 pF]
        ],
        Result->{DB[S21],K},
        Frequency->{1,5} GHz, Step->{1,5}]
```

Out:
```
    Step    DB[S21]   K

    1.      -5.52     1.5

    5.      -8.44     1.5
```

Active networks can have just about any stability factor:

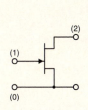

In:
```
NodalAnalyze[
        NodalNetwork[
                MESFET[{1,0,2},4,300,
                        0.3 pF,0.02 pF,0.03 pF,
                        0.04, 2 pSec]
        ],
        Result->{DB[S21],K},
        Frequency->{2,10,18,26} GHz,
        Step->{2,10,18,26}]
```

Out:
```
    Step    DB[S21]   K

    2.      10.5      0.0842

    10.     6.78      0.416

    18.     3.14      0.731

    26.     0.395     1.02
```

Notice that this simple FET model has a stability performance similar to the measured data we used at the beginning of Section 9.2.1.

9.3 Stability Circles

The stability factor, K, indicates if a passive load can make a device unstable, but it gives no information on the load values that would cause the instability. You can use the device's S-parameters to solve for those values of source impedance that cause the output reflection to have a magnitude of unity. Reflection coefficients with magnitudes greater than unity represent negative resistances — symptoms of unstable circuits — which amplifier designers normally avoid. The unity reflection-coefficient magnitude represents lossless circuits, which are the border between lossy and potentially regenerative networks.

The source reflection coefficient is related to the device output reflection coefficient by $\Gamma_{Out} = S_{22} + S_{21} S_{12} \Gamma_S/(1 - S_{11} \Gamma_S)$, and the load reflection coefficient is related to the input reflection coefficient by $\Gamma_{In} = S_{11} + S_{21} S_{12} \Gamma_L/(1 - S_{22} \Gamma_L)$. If you find those values of Γ_S that make Γ_{Out} unity magnitude, you will have defined a circle. The edge of the circle indicates those Γ_S values that make Γ_{Out} unity magnitude.

Formulae for stability circles exist in many textbooks [Gonzalez84], and we can express them as *Mathematica* functions:

```
In:
StabilityCenter[{{s11_,s12_},{s21_,s22_}},
                                       delta_] :=
            Conjugate[s11 - delta Conjugate[s22]]/
                 (Abs[s11]^2 - Abs[delta]^2)

StabilityRadius[{{s11_,s12_},{s21_,s22_}},
                                       delta_] :=
            Abs[s21 s12/(Abs[s11]^2 - Abs[delta]^2)]

StabilityCircles[{{s11_,s12_},{s21_,s22_}}]:=
    Module[ {delta},
            delta = s11 s22 - s21 s12;
            {{cIn -> StabilityCenter[{{s11,s12},
                              {s21,s22}},delta],
              rIn -> StabilityRadius[{{s11,s12},
                              {s21,s22}},delta]},
            {cOut -> StabilityCenter[{{s22,s21},
                              {s12,s11}},delta],
              rOut -> StabilityRadius[{{s22,s21},
                              {s12,s11}},delta]}}
    ]
```

The **StabilityCircles** function uses rules to output data. Rules allow you to identify and clearly extract the data you need. You create the following stability circles by mapping the **StabilityCircles** function through the **sMatrices** data of Section 9.2.1:

```
In:
  sCircs = Map[ StabilityCircles, sMatrices ];
  outCircs = {cOut,rOut} /. Transpose[sCircs][[2]]
Out:
  {{-9.49327 + 205.42 I, 205.544},

   {0.318557 + 3.93119 I, 3.55392},

   {-0.243883 + 3.28138 I, 2.78394},

   {-0.836486 + 2.87917 I, 2.37395},

   {-0.929008 + 2.57812 I, 1.95183},

   {-0.737651 + 2.12929 I, 1.29992},

   {-0.907415 + 1.69164 I, 0.851214}}
```

Next, we plot the output stability circles on a Smith chart. Load reflections within these circles create a negative resistance at the device input and, if a negative input resistance dominates the source resistance, an oscillation will occur. Load reflections outside the stability circle leave the device input impedance positive.

You can use **Map** to create circles with the **outCircs** data, but several problems arise when you create the graphic: The first data point produces a PostScript error in *Mathematica* 2.0, the **PlotLabel** option prints over the lines, and labels are not on the stability circles. You can overcome these problems by using **Show** to drop the first data point and by using **Text** to produce the labels:

```
In:
  Show[smithG,
      Graphics[
          Map[Circle[{Re[#[[1]]],
                Im[#[[1]]]},#[[2]]] ]&,
                Drop[outCircs,1]]
      ],
      Graphics[{
          Text["Output Stability Circles 6-26 GHz",
                {0,-1.25}],
          Text["26 GHz", {-0.5,1.5}],
          Text["6 GHz", {1.75,0.35}]
      }],
      AspectRatio->Automatic,
      PlotRange->{{-2,2},{-1.5,2.5}}];
```

Out:

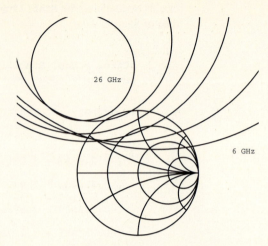

26 GHz

6 GHz

Output Stability Circles 6-26 GHz

You must make sure the output termination for the NEC202 does not lie within the stability circles you just graphed. You should always question if the inside or the outside of a stability circle is the stable region. The key comes from knowing the S-parameters of your device. If the magnitudes of S_{11} and S_{22} are less than unity, the center of the Smith chart is in the stable region; if the magnitude of S_{11} is greater than unity, then load terminations of 50 Ω and the center of the Smith chart are in the unstable region.

Once you know what source and load terminations to avoid, you can proceed with finding the terminations you need for best gain, noise, and (possibly) output power.

9.3.1 Gain Circles

Device S-parameters also are used to calculate device gain. Three gain definitions are popular these days: maximum available gain (MAG), maximum unilateral gain (MUG), and maximum stable gain (MSG) [Gonzalez84]. MAG is also called the *maximum transducer gain*. You can write formulae that calculate device gains as *Mathematica* functions:

In:
```
MAG[s21m_,s12m_,k_] :=
    If[ Abs[k] >= 1,
        N[ Abs[s21m/s12m] (k-Sqrt[k^2-1]) ],
        Infinity
    ]


SetAttributes[MAG,Listable];
```

You can use **SetAttributes**, as we did with **MAG**, to make *Mathematica* map your function over arguments that are lists. You should use **SetAttributes** to make **MAG** listable because the **If** statement within **MAG** expects to test single values of **k**. If you do not make **MAG** listable then you can either use **MAG** only for single sets of parameters or use **Apply** to feed sets of parameters to **MAG**:

```
In:
  MUG[s21m_,s11m_,s22m_] :=
            s21m^2/((1-s11m^2)(1-s22m^2))


  MSG[s21m_,s12m_] := Abs[s21m/s12m]
```

Now that you have functions that evaluate gain, you can tabulate the gains using the S-parameters from the NEC202:

```
In:
  TableForm[
      Transpose[{fs,MSG[s21m,s12m],
            MUG[s21m,s11m,s22m],MAG[s21m,s12m,ks]}],
            TableSpacing->{0,3},
            TableHeadings->{None,
                              {"GHz","MSG","MUG","MAG"}}]
```

```
Out:
      GHz    MSG        MUG        MAG
      2      115.667    1097.93    Infinity
      6      51.8333    104.41     Infinity
      10     31.        42.2707    Infinity
      14     22.9091    23.6255    Infinity
      18     20.3636    14.9382    Infinity
      22     20.1       10.7239    Infinity
      26     17.7       6.8782     10.2747
```

The final amplifier gain will always be less than the available gain, **MAG**. When **MAG** is infinity the amplifier gain must also be below the maximum stable gain or your amplifier will oscillate. **MAG** is infinite for most of the frequencies in the preceding table because the *K* factor is less than unity at those frequencies. The MSG and MUG results show how the maximum gain reduces as frequency increases. For a flat amplifier response over a bandwidth wider than 10%, your design can use suboptimal matching at lower frequencies. Suboptimal matching will reduce the amplifier gain by changing the terminations away from the maximum available gain. Circles of constant gain show you what combinations of source and load impedances give you the gain you need.

You can create many different gain circles. Constant gain circles may be based on source impedances or load impedances, or both. When *K* is less than unity (conditional stability), gain circles cannot be based on a complete conjugate match [Gonzalez84]. When your device is conditionally stable the power gain,

G_p, is the most useful gain measure. G_p is based on establishing an output termination, Γ_L, and then conjugate matching the device input according to Γ_{In}. G_p and g_p, the normalized power gain, are defined by

$$G_p = \frac{1}{1 - |\Gamma_{In}|^2} \, |S_{21}|^2 \, \frac{1 - |\Gamma_L|^2}{|1 - S_{22}\Gamma_L|^2} \quad,$$

$$G_p = g_p \, |S_{21}|^2 \,.$$

The *Mathematica* function **GainCircle** calculates the position and size of a power gain circle and depends directly on G_p and the device S-parameters. You can use the *Mathematica* **Message** function to alert users when they request a G_p larger than *MAG*:

In:

```
GainCircle::usage =
"GainCircle[Gp,{{s11,s12},{s21,s22}}] returns
the gain circle center and radius for a power
gain of Gp.";

GainCircle::gain = "Requested gain is larger than
maximum available gain, `1`";

GainCircle[Gp_,{{s11_,s12_},{s21_,s22_}}] :=
    Module[ {delta,c2,k,c,r,den,lg,gp},
        k = StabilityK[{{s11,s12},{s21,s22}}];
        If[ Gp > MAG[Abs[s21],Abs[s12],k],
            Message[GainCircle::gain,
                    MAG[Abs[s21],Abs[s12],k]];
            Return[$Failed]
    ];
    gp = Gp/Abs[s21]^2;
    delta = s11 s22 - s21 s12;
    c2c=Conjugate[s22 - delta Conjugate[s11]];
    den = (1+gp (Abs[s22]^2-Abs[delta]^2));
    c = gp c2c/den;
    lg = Abs[s12 s21];
    r=Sqrt[1 - 2 k lg gp +(gp lg)^2]/Abs[den];
    N[{cOut->gp c2c/den, rOut->r}]
    ]
```

9.3.2 A Gain Circle Example

As an example, we can define a design specification for 12 dB of gain at 18 GHz. The fifth element of the array **sMatrices** contains the S-parameters at 18 GHz for the NEC202. You can use **Show**, along with our **smithG** graphic, to plot the constant gain circles in the reflection plane:

In:
```
gCirc=GainCircle[ 10^(12/10),sMatrices[[5]] ];

Show[smithG,
    Graphics[
        {Circle[{Re[cOut],Im[cOut]},rOut] /.gCirc}
    ],
    PlotLabel->"Gain Circle for 18 GHz & 12 dB",
    AspectRatio->Automatic,
    PlotRange->{{-1.5,1.5},{-1.0,1.5}}];
```

Out:

Once you pick a point on the gain circle, you can compute the required input termination for a conjugate match. Try picking a point as near as possible to the center of the chart, say at $-0.037 + I\,0.258$. (If you have a Notebook front end to *Mathematica*, you can click on the point you want, copy it, and paste the value into a cell.) You can compute Γ_{In}, given in Section 9.2.3, as follows:

In:
```
gammaL = -0.037 + I 0.258;
{{s11,s12},{s21,s22}} = sMatrices[[5]];
s11 + s21 s12 gammaL/(1 - s22 gammaL)
```

Out:
```
-0.533227 - 0.566026 I
```

You should verify that the conjugate of this Γ_{In}, our desired source termination, lies outside the input stability circle:

In:
```
circ18 = StabilityCircles[sMatrices[[5]] ];
Show[smithG,
        Graphics[ {Circle[{Re[cIn],Im[cIn]},
                              rIn] /. circ18[[1]]}
        ],
        PlotLabel->"InputStab Circle for 18 GHz",
        AspectRatio->Automatic,
        PlotRange->{{-1.5,1.5},{-1.0,1.5}}];
```

Out:

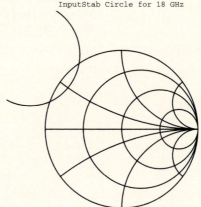

The computed source impedance lies outside the input stability circle, so you can proceed with your design. If you needed to, you could compute constant noise-figure circles to help optimize the amplifier network impedances [Gonzalez84]. For now, we shall stay with gain analysis and leave a discussion of noise figure calculations for Chapter 10.

Next, you must design matching networks to transform 50 Ω source and load impedances to the desired points found using the gain circle technique. Section 9.4 illustrates an empirical approach to matching-network design for narrowband amplifiers. If you need to design broadband amplifiers, you can use feedback or matching networks based on the filter theory of Chapter 8. Two approximations dominate amplifier matching-network realization: First, you cannot fit the gain curves exactly; Second, the device's S_{12} will create interactions between the input and output networks that will change the gain. In the end, you must optimize your amplifier matching networks with a simulator. Even when you have an optimizer, you should start with a good design!

9.4 Matching-Network Design

We shall use the Smith chart mainly as a design aid in this section on matching-network design. Once we plot impedance values on the Smith chart, the lines of constant resistance and conductance, which we described in Section 9.1, will help us to know which types of components are useful for matching and what their values should be: Series reactances move the impedance along lines of constant resistance, whereas shunt susceptances move the impedance along lines of constant conductance. Using a Smith chart for matching network design is like solving a puzzle where you must move a set of dots — the original impedance values — to the center of the chart. By trial and error, you can guess component values, add the components to your circuit, and plot the new impedance values on the Smith chart.

In the past, engineers plotted lines on a Smith chart with a ruler, a compass, and a pencil. But you can build a *Mathematica* function to compute points for your Smith chart traces. You can then use the **SmithCharter** function from Section 9.1 to draw your Smith chart traces.

9.4.1 Smith Chart Impedance Traces

The **smithTrace** function takes an impedance value and a component value, and returns a list of points for plotting. The **smithTraceZ** function will also return a value for the impedance of our new network. Five points are used for the impedance trace and, for now, **zOld** is an arbitrary starting impedance:

```
In:
  za = 25 I;
  zOld = 30 - 10 I;
  zps = za Range[1,5]/5.0 + zOld
Out:
  {30 - 5. I,30 + 0. I,30 + 5. I,30 + 10. I,30 + 15. I}
```

Next, you can transform your new impedance values to the reflection plane. Coordinates in the reflection plane come from the real and imaginary parts of the reflection coefficients. We use a default value of five points in **smithTraceZ**, but you can increase the number of points to improve plot traces:

```
In:
  rhos = rho[zps]
```

Out:
```
{-0.245136 - 0.077821 I, -0.25 + 0. I,

  -0.245136 + 0.077821 I, -0.230769 + 0.153846 I,

  -0.207547 + 0.226415 I}
```

In:
```
smithTraceZ[za_,zOld_,pts_:5] :=
        Module[ {zps,rhos},
            zps = N[ za Range[1,pts]/pts + zOld];
            rhos = (zps - 50.0)/(zps + 50.0);
            {rhos,za+zOld}
        ]
```

Create impedance values between zOld and za+zOld. Convert to reflection values. Return reflections and last total Z.

It is important that you understand how components move impedances on the Smith chart. This graphic shows traces for 40 $I\,\Omega$, −20 $I\,\Omega$, and 30 Ω added to an original impedance of $(35 − 5\,I)\,\Omega$:

In:
```
SmithCharter[
        smithTraceZ[ 40 I, 35 - 5 I ][[1]],
        smithTraceZ[ -20 I, 35 - 5 I ][[1]],
        smithTraceZ[ 30, 35 - 5 I ][[1]]
    ];
```

Out:

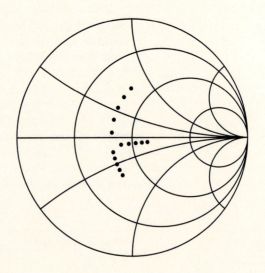

Adding a 40 I Ω inductor to $(35 - 5\ I)$ Ω moves the reflection coefficient up and to the right, along the 35 Ω constant-resistance curve. The $-20\ I$ Ω reactance is a capacitor that moves the reflection down and to the right. A 30 Ω resistor moves the impedance along the $-5\ I$ Ω constant-reactance line toward the center of the chart. (The different element values create unique trace lengths for easier identification.)

9.4.2 Smith Chart Admittance Traces

You must also consider how shunt elements change the impedances. In this section, we describe a function for tracing shunt elements, **smithTraceY** You might also use transmission lines and tapped coils for building matching networks, but a discussion of these is beyond the scope of this book. (There are many good references for matching network design [Bahl88, Ha81, Vendelin90].)

Create a range of admittances between 1/zOld and 1/zOld + 1/za. Then convert these admittances to reflections. Finally return the rhos and final impedance.

```
In:
smithTraceY[za_,zOld_,pts_:5] :=
    Module[ {yps,rhos},
            yps = 1.0/zOld + Range[1,pts]/(pts za);
            rhos = (1/yps - 50.0)/(1/yps + 50.0);
            {rhos, (za zOld)/(za+zOld)}
        ]
```

You can also use **SmithCharter** to examine shunt networks. A shunt inductive reactance of 80 I Ω moves the impedance up and to the left. A shunt capacitive reactance of $-120\ I$ Ω moves the impedance down and to the left. A shunt resistance of 160 Ω moves the impedance to the left. Two important items should be remembered: Lower impedances cause greater changes when in shunt, and shunt networks move along constant conductance and susceptance curves. If you swap the left and right ends of a Smith chart, you create an admittance chart. Shunt networks follow the lines of an admittance chart just as series networks follow the lines of an impedance chart. Overlay charts include both impedance and admittance chart overlaid in red and green ink so the user can easily move series or shunt networks on the same chart.

```
In:
SmithCharter[
        smithTraceY[ 30 I, 35 - 5 I][[1]],
        smithTraceY[ -60 I, 35 - 5 I][[1]],
        smithTraceY[ 80, 35 - 5 I][[1]]
    ];
```

Out:

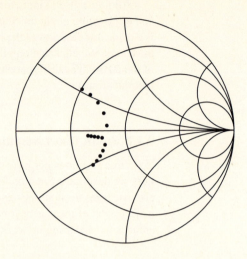

9.4.3 Matching-Network Design

Now we can match our device at a single frequency. Your input network creates a conjugate match by moving Γ_{In} to 50 Ω. Your output network must move 50 Ω to Γ_L. For your input, you start with the required Γ_{In} and imagine what combination of components will move Γ_{In} to the center of the chart. The following Smith chart shows the Γ_{In} point:

In:
```
SmithCharter[ {-0.5332 - I 0.566} ];
```

Out:

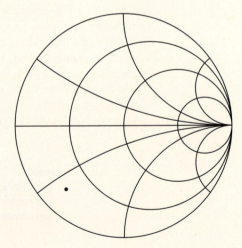

Based on your previous work with a Smith chart, you could use a series resistor to move to the 50 Ω constant resistance circle, and a series inductor to finish the match (shown by the following *Mathematica* code), but this would create a lossy network and reduce amplifier gain. You can now see why the **smithTraceZ** function returns two items (the first item is the list of reflection coefficients; the second item is the final impedance value): Each new component calculation needs the resulting input impedance from the previous component.

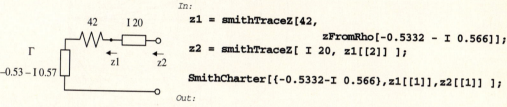

In:
```
z1 = smithTraceZ[42,
                        zFromRho[-0.5332 - I 0.566]];
z2 = smithTraceZ[ I 20, z1[[2]] ];

SmithCharter[{-0.5332-I 0.566},z1[[1]],z2[[1]] ];
```
Out:

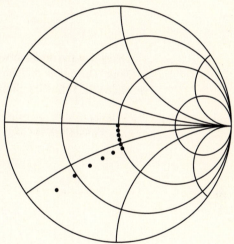

A better design, shown in the following code example, would use a series inductance and shunt capacitance to bring the input impedance to 50 Ω:

In:
```
z1 = smithTraceZ[ I 39,
                        zFromRho[-0.5332 - I 0.566]];
z2 = smithTraceY[ -I 21, z1[[2]] ];
SmithCharter[
        {-0.5332 - I 0.566},
        z1[[1]],
        z2[[1]],
        PlotLabel->"18 GHz Input Matching Network"
    ];
```

Out:

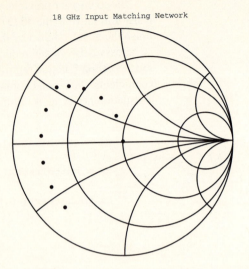

18 GHz Input Matching Network

Now that the input network is done, you can work on the output matching network. The output matching network must bring 50 Ω to Γ_L. The Γ_L point is shown on the following Smith chart:

In:
    SmithCharter[{-0.0367 + I 0.258}];

Out:

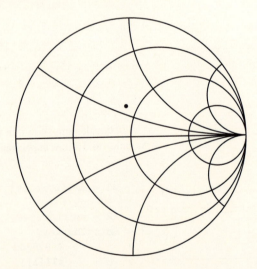

You can use just a shunt inductor to approximate the output matching network. A small series inductor could be used as well to make the match perfect.

In:
```
z1 = smithTraceY[ I 100, 50];
SmithCharter[{-0.0367 + I 0.258},
    z1[[1]],
    PlotLabel->"18 GHz Output Matching Network"
];
```

Out:

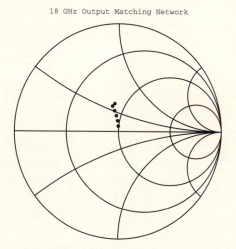

Once the networks are designed, you should simulate the entire amplifier.

9.4.4 Design Evaluation with *Nodal*

You can design matching networks with just a Smith chart and some impedance values. However, you need a CAD program to evaluate your final amplifier at many frequencies. You can use *Nodal* to analyze your final design, to compute the out-of-band stability, and to optimize your component values.

First, you must change the reactance values computed in the previous section to component values. Here, we express component values in pF and nH because these units are convenient at microwave frequencies:

In:
```
cIn = 1/(21 2 Pi 18 GHz)/pF //N
```

Out:
```
0.421045
```

In:
```
lIn = 39/(2 Pi 18 GHz)/nH //N
```

Out:
```
0.344836
```

```
In:
  lOut = 100/(2 Pi 18 GHz)/nH //N
```

```
Out:
  0.884194
```

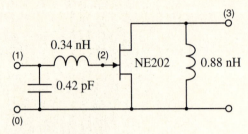

Figure 9.1 NE202 amplifier design.

Your final amplifier circuit, without DC biasing, is shown in Figure 9.1.

Second, you need the **sMatrices** from Section 9.2 converted to a form (that is, Y-parameters) used by the *Nodal* **SubNetwork** function. You can convert the NE202 S-parameter information in **sMatrices** to Y-parameters and append the frequency information:

Label the sMatrices as SParameters and use YParameters to convert the matrices. Append the frequency as a rule.

```
In:
  yLoad[sMatrix_, fr_] :=
      Append[YParameters[SParameters[sMatrix]],
                    Frequency->fr]
```

You can then make a list of Y-parameters for your entire frequency range by using **Apply**. From Section 9.2, the device's S-parameters and frequencies are stored in variables **sMatrices** and **fs**.

Use Transpose to create an array of sMatrices and frequencies. Then use Apply to map yLoad over the array.

```
In:
  yList = Apply[yLoad,
                    Transpose[{sMatrices,fs GHz}],1];

  NodalAnalyze[
          NodalNetwork[
                  Capacitor[{1,0}, cIn pF],
                  Inductor[{1,2}, lIn nH],
                  SubNetwork[{2,3},yList],
                  Inductor[{3,0}, lOut nH]
          ],
          Nodes->{1,3},
          Result->{DB[S11],DB[S21],DB[S22],K},
          Frequency->fs GHz,
          Step->fs
      ]
```

Out:

Step	DB[S11]	DB[S21]	DB[S22]	K
2.	0.105	−1.26	−0.173	0.0931
6.	0.0273	4.57	−1.53	0.271
10.	−0.184	6.25	−3.3	0.373
14.	−1.44	9.36	−2.93	0.496
18.	−30.8	11.9	−2.75	0.692
22.	−2.33	6.14	−11.2	0.919
26.	−0.541	−1.41	−11.8	1.15

Your circuit is very close to the design specification at 18 GHz: The gain is 12 dB, and the input is nearly matched. A perfect match would have **DB[S11]** = − ∞. In practice, any **DB[S11]** less than −20 dB is an excellent match. However, the output match is poor at 18 GHz: Remember that the output was intentionally mismatched to create a stable gain. The *K* values are the same as the device by itself because, since the matching networks are lossless, the *K* factors should not change. In practice, the input and output networks will have loss; small losses in the input network of the NE202 design make the amplifier stable enough to use.

9.5 System Design

High-frequency design is not limited to components. Ultimately, the components must be used in a system; which can be a radio, television, VCR, cellular telephone, or radar — to name a few. Most systems have many components, and each component must have a detailed design before it can be produced. The previous sections focused on that detailed component design.

The first step of system design is a high-level analysis. High-level analysis considers component gains, noise figures, and intercept points. Noise figure will be discussed more in Chapter 10, and we have already discussed gains. An intercept point is a measure of component nonlinearity [Ha81,Tsui83]. So the typical system specification covers gain versus frequency, noise added by the system, and the system nonlinearity. Components must be combined so that the overall system gain, noise figure, and intercept point meet the system requirements. If readily available components can be combined to meet your system needs, then you can buy or build your system.

System engineers typically have a software program to cascade components and predict system specifications. Often system engineers use a spreadsheet to cascade component specifications. *Mathematica* has many advantages over a spreadsheet, especially for cascade analysis. In *Mathematica*, you can use rules for default component parameters, set up custom graphics for plotting cascade results, and draw the resulting system schematic. You can also leave a component parameter as a symbol and solve for the parameter value that meets the system requirements. Plots of cascade performance are particularly useful because components having the greatest impact on performance are easily spotted.

9.5.1 Cascade Analysis Mathematics

In this section, you will develop some code for defining and cascading components. For simplicity, you will consider amplifiers only; later you can extend your cascade analysis program to mixers, attenuators, and filters. Because all components have specifications for gain, noise figure, and intercept points, most of the work for extending this cascade analysis to mixers and other components involves coding the component drawing functions.

For now, you will only consider gain, noise figure (NF), and the third-order intercept point (IP3). These three specifications are the most important, and the programming task is easier to manage with just a few specifications. The first thing you need is the mathematics of cascading components [Tsui83].

Component gains are easy. Gains multiply each other in power or add in dB (see Figure 9.2):

$$G_{\text{Total}} = G_1 * G_2$$

or

$$G_{\text{TotaldB}} = G_{1\text{dB}} + G_{2\text{dB}}.$$

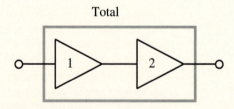

Figure 9.2 Cascade of two amplifiers.

Noise figure cascading works a bit differently. The first-stage gain effectively reduces the second-stage noise contribution. Since all the noise is referred to the input, each component's noise contribution is divided by the total gain preceding the component. Friis' formula, which follows, gives the total noise figure for a cascade of amplifiers [Ambrozy82]. The noise factor, F_i, and gain, G_i, terms in the equation below are power ratios (each *i* subscript stands for a component number). Noise figure, NF, is the decibel representation of the noise factor, *F*.

$$F_{Total} = F_1 + \frac{F_2 - 1}{G_1}$$

Unity is subtracted from F_2 in the above equation because each noise factor contains a contribution from the source noise. Since F_{Total} should count the source noise only once, the source noise is subtracted from every amplifier but the first. Chapter 10 will cover noise figure in more detail.

The procedure for cascading IP3 is more complicated than that for noise figure. Because intercept points may be less familiar to you, we will present a brief explanation [Ha81]. Third-order nonlinearities come from the cubic term in the power series of the amplifier transfer curve. Narrowband systems are mostly concerned with third-order nonlinearities, while broadband systems — such as CATV — are also concerned with second-order nonlinearities. We will assume our signals are small so higher orders can be neglected. The IP3 intercept point is a figure of merit that defines a component's third-order distortion. The idea is that as amplifier output power is increased, the third-order distortion increases three times as fast because of the cubic nonlinearity (see Figure 9.3). At some theoretical output power, the distortion power (d3) and the output power (P_{out}) would be the same. The power level at which output power and distortion power would be the same is the third-order intercept point, IP3. In reality, neither the output power nor the distortion power can reach the intersection point. As a final practical point, IP3 is measured using two drive signals. The drive signals have the same power and are at frequencies f_1 and f_2. The measured distortion will be at either $2f_1 - f_2$ or $2f_2 - f_1$.

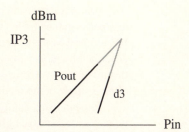

Figure 9.3   Amplifier output power and distortion. The third-order
intercept point is shown.

The previous theoretical background should help you understand the process of cascading amplifier IP3 values. IP3 values are combined in the same manner as shunt resistance values. However, IP3 values are affected by gain. Our cascade program will refer all IP3 values to the cascade input. So each IP3 value will be reduced by the total gain before the amplifier as well as the amplifier gain. IP3 values in data sheets are almost always given at the amplifier output. The equation below defines the total IP3 for the devices in Fig. 9.2:

$$\frac{1}{IP3_{Total}} = \frac{G_1}{IP3_1} + \frac{G_1 G_2}{IP3_2}$$

Now that you have the formulae, we can move on to the program code.

9.5.2 A Cascade-Analysis Program

Your amplifier function will use rules as input. Rules are useful for two reasons: Users do not have to remember argument positions, and users do not have to specify nonessential data. The following code defines the **Amplifier** function and its options. The **Amplifier** options contain default values for the gain, NF, and IP3. The **Amplifier** function consists of evaluating the user rules and the default rules. Note that **Amplifier** inputs are all in dB.

```
In:
    Options[Amplifier] = {Gain->0,NF->0,IP3->99};


    ampEval[model___Rule] :=
      {Gain,NF,IP3} /. {model} /. Options[Amplifier]
```

Actually, **Amplifier** is just a wrapper — the real work is handled by the **ampEval** function. Because you need to tabulate and draw the cascade system, it is better to hold amplifier information in a wrapper than in a list of numbers. The technique of using wrappers will be even more important if you extend this program to other components such as mixers and attenuators.

Your main cascade analysis function will be **Tabulate**, which creates a system performance table. The performance table shows the system gain, NF, and IP3 referred to the system input as each amplifier is added. The previously described formulae compute the system performance. The only remaining subtleties involve converting decibels to the power ratios used in the formulae and handling a single component.

You can use a datatype, **Cascade**, to wrap the components and another datatype, **CascadeForm**, to hold the tabulated results. The **CascadeForm** datatype allows **Format** to display the table in a nice way and allows the plotting code to be sure of its inputs:

Range and Length are used in TableHeadings to label each component by number.

```
In:
    Format[CascadeForm[x_]] ^:=
            TableForm[x,TableSpacing->{1,3},
                TableHeadings->{{"Gain","NF","IP3"},
                    Range[ Length[ Transpose[x] ]]}]
```

```
Normal[CascadeForm[x_]] := x
```

The cascade formulae result in accumulations over many components. The **ListSum** function works as an accumulator:

In:
```
ListSum[x_List] := Rest[FoldList[Plus,0,x]]

Tabulate[Cascade[system__],opts___Rule] :=
   Module[ {tmp,sys,trSys,cGainsDB,gains,
                       cGains,Fs,cNF,cIP},
           sys = {system} /. Amplifier->ampEval;
           If[ Length[{system}] === 1,
               sys = Flatten[sys];
               CascadeForm[ N[{{sys[[1]]},
                 {sys[[2]]},{sys[[3]]-sys[[1]]}},3]
               ],
               trSys = Transpose[sys];
               cGainsDB = ListSum[ trSys[[1]] ];
               gains = 10^(trSys[[1]]/10.0);
               cGains = Drop[10^(cGainsDB/10.0),-1];
               Fs = 10^(trSys[[2]]/10.0);
               tmp = Rest[(Fs-1)]/cGains;
               cNF = 10.0 Log[10, ListSum[
                           Prepend[tmp,Fs[[1]]] ] ];
               tmp = Prepend[cGains,1]/
                     (10^((trSys[[3]]-trSys[[1]])/10.0));
               cIP = -10.0 Log[10,ListSum[tmp]];
               CascadeForm[{cGainsDB, cNF, cIP}]
           ]
       ]
```

Change all the Amplifier wrappers to ampEval. If there is only one component, just create a Cascade-Form with gain, NF, and IP3 referred to the input.

Once you have the **Tabulate** function, you can evaluate a system of any size. As an example, we will examine a system with three identical amplifiers:

In:
```
Amp1 = Amplifier[Gain->10,NF->1.5,IP3->30];

Tabulate[Cascade[Amp1,Amp1,Amp1]]
```
Out:

	1	2	3
Gain	10	20	30
NF	1.5	1.62502	1.63733
IP3	20.	9.58607	-0.45323

Each amplifier is denoted **Amp1**. Each amplifier has 10 dB of gain, a noise figure of 1.5 dB, and an output IP3 of 30 dBm. The system table shows the system input performance as each amplifier is added. The **1** column shows a single amplifier. Remember that IP3 at the amplifier input is the output IP3 minus the device gain in dB. The second column shows two devices. Here we see a small increase in noise figure and an IP3 dominated by the second stage. Adding a third amplifier shows a definite trend in all performance areas.

What if you need to convert a system specification into a component specification? *Mathematica* functions such as **FindRoot** are just what you need. Assume you need a system input IP3 of 3 dB rather than the –0.45 dB previously calculated. We know the last amplifier limits our IP3 because up to our last amplifier the system IP3 is 9.59 dB. What would the IP3 of the last amplifier have to be to meet our new specification?

You can set up an amplifier, **Amp3**, with IP3 as a symbol, **x**. Then you can evaluate your cascade using **Tabulate**:

```
In:
  Amp3 = Amplifier[Gain->10,NF->1.5,IP3->x];


  tbl = Tabulate[Cascade[Amp1,Amp1,Amp3]];
```

You can extract the system IP3 for all the amplifiers as the third value in the third row of **tbl** — once you have removed the **CascadeForm** wrapper using **Normal**:

```
In:
  Normal[tbl][[3,3]]

Out:
                               100.
      -10. Log[0.11 + ---------------]
                          0.1 (-10 + x)
                      10
      ---------------------------------
                  Log[10]
```

The preceding formula for system IP3 can be solved for **x** with **FindRoot**. All you have to do is subtract the system specification, 3 dBm, from the system IP3 formula and give **FindRoot** some starting values for **x**:

```
In:
  FindRoot[ Normal[tbl][[3,3]] - 3,{x,{30,40}}]

Out:
  {x -> 34.0762}
```

An IP3 of 34.1 dBm for the last amplifier will meet our system input IP3 specification of 3 dBm. So you must increase the output amplifier intercept point — and cost. However, our new output amplifier IP3 is only 4.1 dB higher than the IP3 of the old amplifiers, and the change gave us 3.55 dB improvement in our system IP3.

9.5.3 Plotting the Cascade Analysis

Plotting the cascade results often shows the critical components much better than tabulating them. The **PlotCascade** function is defined next. Actually, two functions are defined. The first version of **PlotCascade** recognizes a system cascade and converts it to a table for plotting. The second **PlotCascade** program takes a table of results and does the plotting. So you will overload the definition of **PlotCascade** to make the function easier to use. The plotted lines use different line styles for easier viewing. The plot would be trivial except that you should show the steps in performance as each amplifier is added to the cascade. To create steps in the lines, you double the number of points using the **dup** function. A similar doubling in the part numbers allows **Line** to plot the performance steps.

```
In:
  dup[x_List] := Flatten[Transpose[{x,x}]]

  PlotCascade[x_Cascade,opts___Rule] :=
                    PlotCascade[Tabulate[x],opts]

  PlotCascade[CascadeForm[x_],opts___Rule] :=
     Module[ {parts,g,nf,ip},
          {g,nf,ip} = x;
          parts = Range[Length[g]];
          parts = Flatten[
                    Transpose[{parts,parts+1}]];
          Show[
            Graphics[{
               Line[ Transpose[{parts,dup[g]}] ],
               Dashing[{0.03,0.03}],
               Line[ Transpose[{parts,dup[nf]}] ],
               Dashing[{0.03,0.01,0.005,0.01}],
               Line[ Transpose[{parts,dup[ip]}] ]
            }],
            Axes->Automatic,
            AxesLabel->{"Devices","dB"}
          ]
     ]
```

When you plot the cascade performance and add more amplifiers, the trends are even more obvious. The identical IP3 levels of all the amplifiers allow the output amplifier to dominate. Only a small correction for the preceding stages is notable in the following plot:

In:
```
PlotCascade[Cascade[Amp1,Amp1,Amp1]];
```

Out:

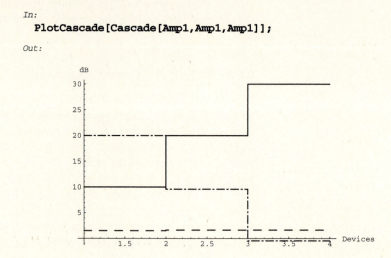

9.5.4 Drawing the Cascade

Finally, you should document the system with a schematic. Drawing a schematic requires three steps: drawing the individual parts, listing the parts, and placing the parts. You can create an amplifier drawing from a **Line** graphic, as shown below. The function **ampDraw** uses coordinates as an input so that the **Draw** function can place many amplifiers. Your list of parts comes from the **Cascade** system description. Finally, the placement is a simple left-to-right ordering; since we are only analyzing cascades, a left-to-right ordering suffices.

In:
```
ampDraw[{xc_,yc_}] :=
        Graphics[{
              Line[{{xc,yc},{xc+0.1,yc},
                     {xc+0.1,yc+0.1},{xc+0.25,yc},
                     {xc+0.35,yc},{xc+0.25,yc},
                     {xc+0.1,yc-0.1},{xc+0.1,yc}}
              ]
        }]
```

The **DrawCascade** function has two parts. First the **DrawCascade** function replaces the **Head** of each **Amplifier** component with **ampDraw** in the cascade. If you make a more general cascade analysis, you can substitute several component drawing functions here. Next, each **ampDraw** function is evaluated at an offset of 0.35 from the previous component. If only one amplifier is found, it is drawn on its own. The resulting set of graphics is sent to **Show** for display.

In:
```
DrawCascade[Cascade[x__],opts___Rule] :=
    Module[ {tmp,sys,pts},
          If[ Length[{x}] === 1, (* 1 in cascade *)
                sys = Head[x][{0,0}] /.
                             Amplifier -> ampDraw;
              , (* else > 1 in cascade *)
                tmp = Map[Head,{x}] /.
                             Amplifier -> ampDraw;
                sys = Table[tmp[[i]][{0.35 i,0}],
                             {i,Length[{x}]}];
          ];
          Show[sys,AspectRatio->Automatic]
    ]
```

The following **Draw** function illustrates our cascade of three amplifiers. In all honesty, a cascade of three amplifiers is not an interesting graphic; but it does emphasize exactly what was analyzed.

In:
```
DrawCascade[Cascade[Amp1,Amp1,Amp1]]
```
Out:

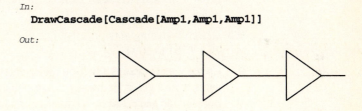

9.6 Summary

In this chapter you learned how, with the help of the Smith chart,

- to design high-frequency circuits using S-parameters
- to link component impedances with S-parameter measurements
- to represent those impedances that made a device unstable or to extract a certain gain from a device
- to design matching networks
- to code a system cascade analysis program in *Mathematica*.

CAD programs can do most of your design by optimization and can check the final design against the original specifications. Some people might be tempted to completely design a circuit by optimization, but optimizers are dangerous because they do not give you an understanding of your design and of the component sensitivities. For example, in your NE202 output matching network, a series inductor was omitted because you could see on the Smith chart that an inductor was not required. The Smith chart has survived because most analog designers need to understand the product they build.

9.7 Exercises

9.1 Create an option for **SmithCharter** to connect the dots with lines.

9.2 Find the K factor of a lossless network.

9.3 Find the K factor of a lossy network.

9.4 Find the K factor of a shunt network of -20 Ω in parallel with 0.25 nH and 1 pF to ground. Place 2 dB and 10 dB pads before and after the circuit, then analyze the circuit from 8 to 13 GHz. What are the circuit pole locations for the two pad configurations? What are the implications of the K factors for the two pad configurations?

9.5 Modify the **smithTrace** functions so the user can specify how many points are used.

9.6 Design matching networks that match $\{25 - I\,25, 25 + I\,25, 75 - I\,75,$ and $75 + I\,75\}$ to 50 Ω.

9.7 Create a **SmithMatch** function that creates traces from a list of components over a given frequency range. This moves the **smithTrace** functions inside **SmithMatch** so the user can have more of a netlist format.

9.8 Analyze the NE202 amplifier design at frequencies near 18 GHz. Determine the -1 dB bandwidth.

9.9 You are given two amplifiers and an attenuator. The attenuator can is variable from 0 to 10 dB and has infinite IP3 (an IP3 of 99 is a good approximation). Amplifier "a" has: NF = 2 dB, Gain = 28 dB, and IP3 = 30 dBm. Amplifier "b" has: NF = 6 dB, Gain = 10 dB and IP3out = 35 dBm. How should you configure the cascade to have maximum $IP3_{in}$, a noise figure less than 6.5 dB, and a gain of 33 dB?

9.10 Modify the cascade analysis functions (**Tabulate**, **PlotCascade**, **DrawCascade**) so that they recognize and use a pad and a mixer.

9.8 References

[Ambrozy82] Ambrozy, A., *Electronic Noise*, McGraw-Hill, New York, 1982.

[Bahl88] Bahl, I., Bhartia, P., *Microwave Solid State Circuit Design*, John Wiley & Sons, New York, 1988.

[Gonzalez84] Gonzalez, G*., Microwave Transistor Amplifiers*, Prentice-Hall, Englewood Cliffs, New Jersey, 1984.

[Ha81] Ha, T., *Solid-State Microwave Amplifier Design*, John Wiley & Sons, New York, 1981.

[Maeder90] Maeder, R., *Programming in Mathematica*, Addison-Wesley, Reading, Massachusetts, 1990.

[Smith39] Smith, P., "Transmission Line Calculator," *Electronics*, 12:29-31, 1939.

[Tsui83] Tsui, J. B., *Microwave Receivers and Related Components*, *NTIS*, PB84-108711, 1983.

[Vendelin90] Vendelin, G., Pavio, A., Rohde, U., *Microwave Circuit Design*, John Wiley & Sons, New York, 1990.

CHAPTER 10

Noise Analysis

Because noise poses a fundamental limit on accurate signal transfer and measurement, noise control is an important consideration in a project's design: The performance of many systems is governed ultimately by a signal-to-noise ratio requirement that translates into a minimum detectable signal (MDS), an acceptable bit error rate (BER), or carrier-to-noise ratio (CNR) [Carlson81]. Shannon first placed communications and noise in its proper perspective of bandwidth and accuracy [Shannon48]. Although at first noise was considered harmful [Johnson71], the profitable use of noise in signal coding for secrecy and signal robustness is becoming standard practice [Pickholtz82].

The analysis of noise in systems is more complicated than signal analysis because you must contend with noise sources throughout a system, consider both the amplitude and the phase disturbances noise causes, and recognize that correlation plays a key role in measuring noise.

In this chapter, we show you how to use *Mathematica* to set up mathematical tools and methods for studying noise. First, we establish the relationships among statistics, time-domain analysis, and frequency-domain analysis. Next, we show how the use of noise correlation leads to the rigorous analysis of noise in linear and quasilinear systems. Quasilinear systems are in a nonlinear operating state but have noise voltages so small they may be treated by linear methods; mixers and oscillators are typical quasilinear systems. Finally, we show how the *Nodal* package solves noise problems.

10.1 Random Signals

If you have a signal measured over all time, you obviously have a complete description of the signal. However, signals cannot be measured over all time, and so a simple description of a signal is necessary to make signal analysis manageable. To describe a noise signal completely, you require a probability density function (PDF) and a spectrum; deterministic signals (for example, a cosine waveform) are fully described by their spectrum.

Because many signals can have the same spectrum, you need to know more about a signal than just its spectrum. The signal statistics, or the fact that the signal is deterministic, is the other information you need to describe your signal.

The PDF, or even the cumulative distribution function (CDF), gives you a compact statistical description.

10.1.1 White Noise

Mathematica is a useful tool for examining several types of noise signals and comparing them with deterministic signals. You can use *Mathematica*'s **Table** function to create a white-noise signal. The following function creates a random number drawn from a Gaussian distribution:

```
In:
   Needs["Statistics`DataManipulation`"]
   Needs["Statistics`NormalDistribution`"]
```

Toolbox

Random

Random[] gives a uniformly distributed pseudorandom real number in the range of 0 to 1.

You can also define the type of random number returned and the range of values from which it is drawn by using **Random[***type*,*range***]**, where *type* is one of **Integer**, **Real**, or **Complex** and *range* is specified as a two-long list {*minValue*, *maxValue*}.

Other distributions (for example, **NormalDistribution**) are available in the statistics packages that come with *Mathematica*.

NormalDistribution

NormalDistribution[*mu*, *sigma***]** generates a random number drawn from a Normal , or Gaussian, distribution with mean *mu* and standard deviation *sigma*.

```
In:
   white=Table[Random[
              NormalDistribution[0,1]],{2048}];
```

```
ListPlot[ Take[white,256],
         PlotJoined->True,
         AxesLabel->{"time","V"},
         PlotLabel->"White Noise"];
```

Out:

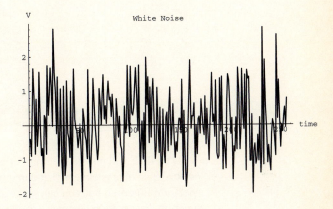

For clarity, only the first 256 points of the white-noise signal are shown. You can calculate the first two moments of the white-noise signal by using functions from the **DataManipulation** package:

In:
```
{Mean[white], Variance[white]}
```

Out:
```
{0.00175441, 1.01035}
```

Despite the short sample length, the mean and variance are close to the theoretically correct values of 0 and 1.

A histogram of sample values shows an approximation of the PDF of your signal. You can generate the histogram by counting your signal with **BinCounts** and displaying the result with **BarChart**. The following bar chart is similar to a Gaussian PDF, but it does not exactly agree because you used a short finite-length sequence.

In:
```
Needs["Graphics`Graphics`"]

BarChart[ BinCounts[white, {-3.0,3.0,0.2}]]
```

Out:

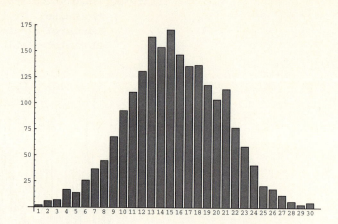

The PDF and power spectrum of a noise signal completely characterize that noise signal. The packages that come with *Mathematica* do not contain a power spectrum function, but you can make a function by using the **InverseFourier** function discussed in Chapter 6. Once you multiply the result of the Fourier transformation by its conjugate, you have the power spectrum. You should use the **Abs** function to force your function to return real numbers. Because the result from the *Mathematica* **InverseFourier** function is not scaled, you should divide the power spectrum by the number of points in the transform.

We use the **InverseFourier** function (rather than **Fourier**) to take a time-domain signal to the frequency domain; engineers and physicists use opposite signs when defining Fourier transforms [Oppenheim75, Wolfram91]:

In:
```
PowerSpectrum[x_List]:=
    Module[{tmp},
            tmp = InverseFourier[x];
            Chop[Abs[tmp Conjugate[tmp]]/
                                Length[x]]
        ]
```

If you plot the power spectrum, you find an essentially flat, but somewhat noisy, spectrum:

In:
```
ListPlot[ DBP[whitePS = PowerSpectrum[white]],
            PlotJoined->True,
            AxesLabel->{"Sample","dB"},
            PlotLabel->"White Power Spectrum"];
```

Out:

Using **Mean**, you can show that the average spectral value is –33.1 dBv:

In:
```
DBP[ Mean[whitePS] ]
```
Out:
```
-33.0707
```

If you sum the noise in all the spectral bins, you should obtain the total signal power, which is equal to the variance of the original signal. You can calculate the total power, **pwr**, by accumulating all the bins in the power spectrum:

In:
```
pwr = Apply[Plus,whitePS]
```
Out:
```
1.00986
```

This number is very close to the noise power computed using the time-domain variance.

Another way to understand the spectrum is to listen: Your ear is one of the best spectrum analyzers made. If your computer supports the sounds facilities in *Mathematica*, you can use the **ListPlay** function to send a signal through your computer's speaker. Notice that the white-noise sound seems random and biased toward high frequencies:

In:
   ListPlay[white];

Out:

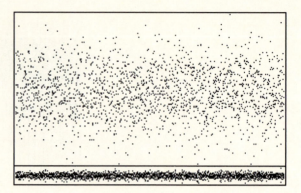

–Sound–

10.1.2 Brown Noise

Brown noise is filtered white noise. Chapter 6 discussed frequency-domain filtering, but, to provide some variety and insight, we work with time-domain filtering in this chapter. The following finite-difference equation creates a first-order low-pass filter. The low-pass filter induces some memory into a signal by adding some of the last output value to the current input value. The dull sound of low-pass filtered noise led to the name brown noise [Gardner78, Keshner82].

$$y(n) = x(n) + a\, y(n - 1)$$

where $y(0) = 0$ and $a < 1$.

You can use **FoldList** and a pure function to convert the white noise into brown noise. For your first example, let a be 0.9. The brown-noise spectrum rolls off as frequency is increased and the histogram of the brown noise is close to that of the white noise:

Slot #1 contains the last (n-1) y value. Slot #2 contains the present (n) x value from white. Drop leaves off the beginning 0.

In:
```
brown = Drop[FoldList[(#2 + 0.9 #1)&,0,white],1];

ListPlot[ Take[brown,256],
          PlotJoined->True,
          AxesLabel->{"time","V"},
          PlotLabel->"Brown Noise"];
```

Out:

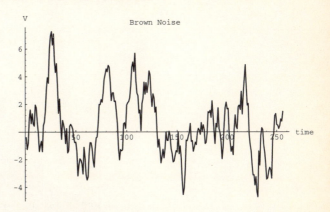

Note how the brown noise appears smoother, and has more low-frequency structure, than the white noise. Again, for clarity, the plot shows only the first 256 signal points.

In:
```
ListPlot[DBP[PowerSpectrum[brown]],
        PlotJoined->True,
        AxesLabel->{"Sample","dB"},
        PlotLabel->"Brown Power Spectrum"];
```

Out:
```
]
```

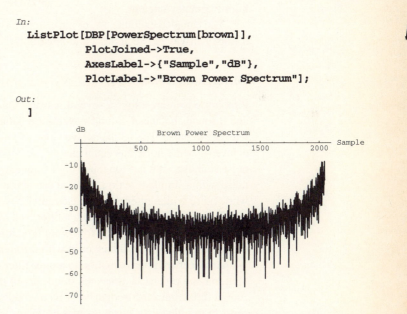

You can see the low-frequency emphasis in the power spectrum of the brown noise. If you smoothed the brown-noise power spectrum, you would see close to a –6 db/octave slope in frequency above the *z*-domain pole at $a = 0.9$.

In:
```
BarChart[ BinCounts[brown, {-6.0,6.0,0.4}]]
```

Out:

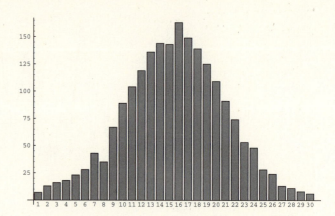

You should also play the brown noise; you will hear a much duller sound than that of the white noise. Depending on your viewpoint, the dull sound is due to too much correlation (produced by the feedback or memory of the filter) or too little high-frequency content.

You can also examine $1/f$, or pink, noise. The pink-noise spectrum will have a slope halfway between the white-noise and brown-noise spectra.

10.1.3 Pink Noise

Pink noise is interesting for several reasons: Qualitatively, it sounds interesting and it sounds balanced. Sound systems are flattened with pink noise rather than with white noise. The –3 dB/octave slope of pink noise creates equal energy per octave — white noise has equal energy per hertz. The flat spectrum of white noise sounds as if it has too much high-frequency content. You cannot create pink noise with a simple function, but you can approximate pink noise by filtering white noise with a filter made by cascading several pole/zero networks [Keshner82, National76]. To make such a filter, you need at least one filter section for each decade of response. The following function approximates pink noise by cascading three first-order filters, each with different time constants.

An excellent approximation to a pink-filter response results when each filter section has a pole at 10 times the frequency of the previous section's pole and a zero at three times the frequency of its own pole. Translating these poles and zeros to sampled data systems gives us values for a sampled data (digital) filter. You can translate the poles from s to z using $z = E^{s\tau}$, where τ is the sampling period. For our purposes, τ can be unity and the s-plane pole/zero locations can range from –0.033 to 10.0. You can create a variable, **zs**, to hold the z-plane root locations:

```
In:
  zs = E^(-{{0.1,0.33},{1.0,3.3},{10.0,33.3}})
```

```
Out:
  {{0.904837, 0.718924}, {0.367879, 0.0368832},
                                          -15
       {0.0000453999, 3.45139 10   }}
```

Each pair of roots forms a filter section for the data. By combining the poles and zeros, you can build a difference formula for the approximate pink filter [Oppenheim75]:

$$y(n) = x(n) + a_1 \, y(n-1) + a_2 \, y(n-2) + a_3 \, y(n-3) +$$
$$b_1 \, x(n-1) + b_2 \, x(n-2) + b_3 \, x(n-3)$$

where $y(0) = 0$ and $a < 1$.

Each of the y-coefficients in the difference equation comes from a combination of the z-plane pole locations. You can use **Expand** to deduce how the pole locations combine to create the difference equation coefficients:

Start the data with a 0.
Map the zero through the x data by taking pairs of adjacent points.
Use the memory of Fold-List to create the pole filter.

```
In:
  zFilterPZ[data_,{p_,z_}] :=
     Module[{xTmp},
        xTmp = Join[{0},data];
        xTmp = Map[({-z,1.0} . #)&,
                          Partition[xTmp,2,1]];
        FoldList[(#2 + p #1)&,0,xTmp]
     ]
```

```
In:
  pink = zFilterPZ[
             zFilterPZ[
               zFilterPZ[white,zs[[1]]],
               zs[[2]] ],
             zs[[3]] ];
```

The following pink-noise plot shows fewer zero-crossings than that of white noise and yet is not as smooth as that of brown noise. The pink-noise waveform is typical of many natural processes (for example, oscillations of crystals and heartbeats [D'Amico86]).

```
In:
  ListPlot[ Take[pink,256],
            PlotJoined->True,
            AxesLabel->{"time","V"},
            PlotLabel->"Pink Noise"];
```

Out:

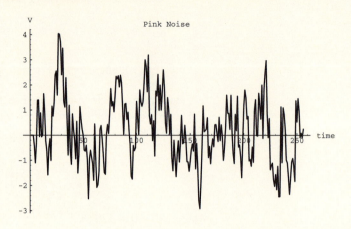

In:

```
ListPlot[DBP[PowerSpectrum[pink]],
        PlotJoined->True,
        AxesLabel->{"Sample","dB"},
        PlotLabel->"Pink Power Spectrum"];
```

Out:

Although the data are noisy, you can see a –10 dB per decade (approximately) slope in pink noise. At the least, the pink-noise slope is approximately halfway between the slopes of white noise and brown noise.

In:
   BarChart[BinCounts[pink, {-4.0,4.0,0.3}]];

Out:

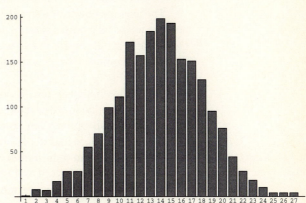

If you compare the histograms of white, brown, and pink noise signals, you can see that they are independent of the noise spectrum and so cannot serve to distinguish the different types of noise. If you want a time-domain tool to distinguish between signals, you can use correlation: Correlation is really just viewing a spectrum in the time domain. Using *Mathematica*, you can correlate a signal with itself (autocorrelation) or with another signal (crosscorrelation).

Through correlation, you can trace the origins of a noise signal. In Section 10.2, you will correlate some of the noise signals we just created. Signals that are related to each other, such as **white** and **brown** in this section, will show some correlation. Unrelated signals will not show any correlation — within the accuracy of our analysis.

Representing correlation as a function of frequency will be much more useful when working with frequency-domain circuit analysis. In fact, noise sources and their correlations are intimately related to component values and network responses.

10.2 Autocorrelation and Power

A sinusoidal signal is predictable; noise is not. You can fully characterize a sinusoid by knowing its period or frequency, its magnitude, and its phase relative to a reference point in time: The Fourier transform of a sinusoid yields a complex number at a frequency. The Fourier transform of a short-duration sample of noise is still noise, so we must use other methods to characterize noise. We cannot predict what value a noise voltage will have at any point in time — the only characteristics we can calculate relate to the mean value, mean square value, and other moments of the noise. Fortunately, these statistical quantities in the time domain relate to the DC and AC power of the signal in the frequency domain — so you can use statistics to give numbers that represent the noise signal.

The average power of a signal at various frequencies, as well as at DC, can be computed as the power spectral density (PSD) of the signal. You can compute the power spectral density in two ways: by multiplying the Fourier transform by its complex conjugate or by taking the Fourier transform of the signal's autocorrelation function.

Correlation has a key role in evaluating the relationships between noise sources in networks, so we develop a correlation and an autocorrelation function. You can set up a function, **Correlation**, to take one or two arguments (that is, overload the function) and so perform either autocorrelation or crosscorrelation. The Fourier transform of the crosscorrelation is the cross-spectral density. The cross-spectral density is very useful when relating noise sources within a network. A **CrossPowerSpectrum** function is defined below. The cross-spectral density can be complex, so we do not use the **Re** function to force the function values to be real numbers. Note that you could define the **CrossPowerSpectrum** function by overloading the **PowerSpectrum** function. You also need a second signal, **white2**, for correlation studies.

```
In:
  white2 = Table[
            Random[NormalDistribution[[0,1]],
            {2048}];

Correlate[x1_List] :=
   N[Sqrt[Length[x1]]] *
        Fourier[ CrossPowerSpectrum[x1,x1] ]

Correlate[x1_List,x2_List] :=
   N[Sqrt[Length[x1]]] *
        Fourier[ CrossPowerSpectrum[x1,x2] ]

CrossPowerSpectrum[x1_List,x2_List] :=
   Module[{tmp,tmp2},
        tmp = InverseFourier[x1];
        tmp2 = InverseFourier[x2];
        Chop[tmp Conjugate[tmp2]/Length[x1]]
   ]
```

You can see from the following plots how autocorrelation produces a spike at zero frequency (DC). The zero-frequency value of the autocorrelation is the variance of the signal. In comparison, the crosscorrelation between signals **white** and **white2** shows only noise:

In:
```
ListPlot[ Take[cor1=Re[Correlate[white]],256],
          PlotJoined->True,
          AxesLabel->{"time","C"},
          PlotLabel->"Autocorrelation",
          PlotRange->All];
```

Out:

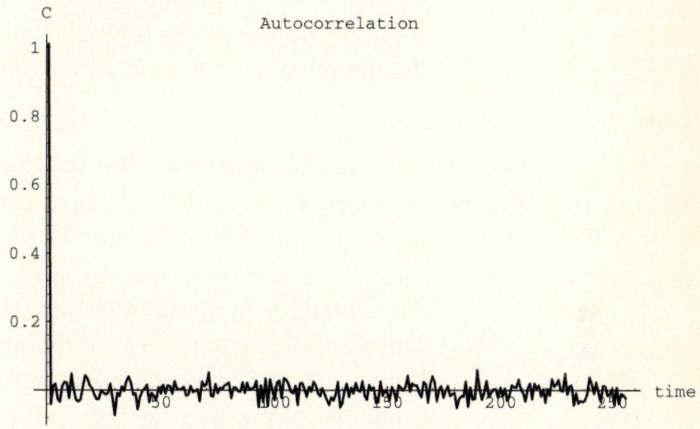

In:
```
ListPlot[ Take[ cor12 =
               Re[Correlate[white,white2]],256],
          PlotJoined->True,
          AxesLabel->{"time","C"},
          PlotLabel->"Crosscorrelation",
          PlotRange->All];
```

Out:

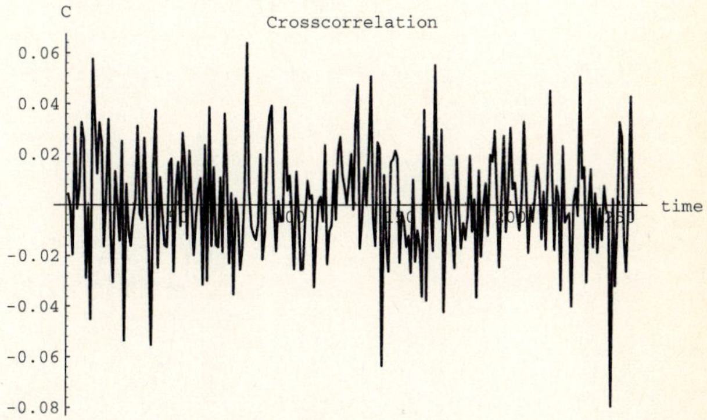

Once again, only the first 256 signal points are shown.

The total power in each correlation gives a measure of the correlation. You obtain the total power in each correlation from the first point of the autocorrelation or the maximum of the crosscorrelation. You need to find the maximum in the crosscorrelation because time shifts between data sets move the maximum away from the zero time point. The following total power calculations show how the white-noise power is equal to the variance. The total power computed by the autocorrelation function is exactly equal to the total power computed by summing the noise in each frequency bin, **pwr**. The crosscorrelation power is small compared to the power in either signal, indicating little correlation.

In:
```
{First[cor1],Max[cor12]}
```

Out:
```
{1.00986, 0.071805}
```

Finally, you should consider what happens when you compare filtered noise with itself. If there is no correlation between noise sources, you can add their powers. However, when noise sources are correlated they may add, subtract, or combine somewhere between complete addition or subtraction. By correlating the white and brown noises you created, you can get a better feel for working with noise. Remember, in our example the brown noise is a filtered version of the white noise, so we should expect some correlation.

In:
```
ListPlot[Take[Re[Correlate[brown,white]],256],
        PlotJoined->True,
        AxesLabel->{"time","C"},
        PlotLabel->"w-b Crosscorrelation",
        PlotRange->All];
```

Out:

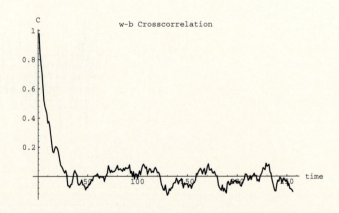

The crosscorrelation between the white and brown noise approximately recovers the brown-noise filter impulse response. The short sequences and finite lengths do add some errors, but the correlation technique can be used for system identification [Oppenheim75].

For our purposes, the power spectrum and the cross-power spectrum of noise will be most important because our noise analysis will operate in the frequency domain. The following plot shows a cross-power spectrum of white and brown noise. Although the following example is too noisy for accurate work, the brown filter response can be seen. (You can use a smoothing function to make the filter response clearer.)

In:
```
ListPlot[DBP[CrossPowerSpectrum[white,brown]],
        PlotJoined->True,
        AxesLabel->{"Sample","dB"},
        PlotLabel->"Cross-Power Spectrum"];
```

Out:

10.3 Multiple Signals and Correlation Matrices

All circuits contain many sources of noise. Each resistor in a circuit has an associated thermal-noise voltage. Each transistor in a circuit contains thermal, shot, and $1/f$ noise sources. When you analyze a circuit, you must sum up the contributions of all the noise sources. One approach is to analyze the circuit response to each noise source and add up the resulting output powers. This approach results in many separate analyses. To make matters worse, any change in source impedance requires the noise analyses to be completely redone. A better way to analyze circuit noise is with correlation matrices [Haus59, Hillbrand76, Ambrozy82].

Correlation matrices work much like circuit matrices: An equivalent response is associated with each network port. In circuit matrix analysis, the transfer function between each pair of ports was stored in a matrix. In circuit noise matrix analysis, each network port has an associated noise source that contains a combination of the circuit's internal noise sources. Thus a large number of internal noise sources are reduced to just a few equivalent port noise sources. Just as in Sections 10.1 and 10.2, you use correlations and spectral densities of noise sources, rather than an infinite set of random values, as simple descriptions of the sources' outputs. The next section discusses correlation matrices, which are a convenient representation of the spectral densities of network noise sources. Section 10.3.2 reviews thermal noise and Section 10.3.3 ties together circuit noise sources, network port noise sources, and correlation matrices.

10.3.1 The Correlation Matrix

This section sets up the mathematics of correlation matrices through *Mathematica* functions. Your first step is to assume that you have many noise signals: say, **n1**, **n2**, **n3**. Creating a vector of noise signals is the easiest way to analyze multiple noise signals. You will need to compute the autocorrelations and crosscorrelations between all the noise signals and you can create the required combinations of noise signals using the *Mathematica* function **Outer**.

We use the **Outer** function with **Cor** to show how you can create a correlation matrix from the noise vector **{n1,n2,n3}**; **Cor** is a dummy function to represent correlation. Once you compute the correlation or power spectrum of the noise-signal combinations, you have a complete picture of the interaction between all the noise sources:

```
In:
  Outer[Cor, {n1,n2,n3}, {n1,n2,n3}]//MatrixForm

Out:
  Cor[n1, n1]   Cor[n1, n2]   Cor[n1, n3]
  Cor[n2, n1]   Cor[n2, n2]   Cor[n2, n3]
  Cor[n3, n1]   Cor[n3, n2]   Cor[n3, n3]
```

The matrix created by **Outer** is called a *correlation matrix*. The **Cor** function, used as a wrapper in the previous calculation, could be a correlation or power spectrum function; since we will work in the frequency domain, we will use power spectra in this chapter.

10.3.2 Resistor Noise

There are many types of noise in circuits: thermal noise, shot noise, trap noise, $1/f$ noise, and others [Ambrozy82]. Fundamental noise sources tend to come from thermal agitation in conductive media (thermal) or discrete charge crossing a potential barrier (shot). We will focus on thermal noise sources in this chapter, although the methods we develop work for all noise sources.

Thermal noise in resistors is essentially white Gaussian noise. The white-noise approximation holds for frequencies up to hundreds of gigahertz. (Depending on the temperature, quantum mechanical considerations change thermal noise sources around 100 GHz [Ambrozy82].) So the thermal noise of the resistor in Figure 10.1 behaves like the white noise discussed in Section 10.1. Resistor noise has a delta function for its autocorrelation and a flat power spectrum.

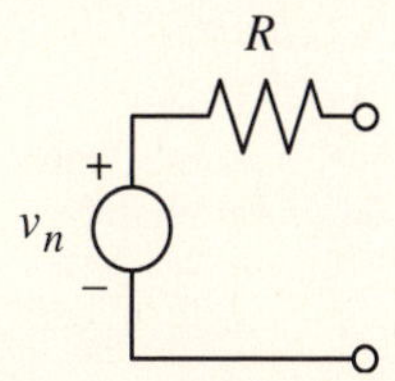

Figure 10.1 Resistor with internal noise voltage.

The power available, P_{av}, from a resistor depends on the temperature and the measurement bandwidth as shown by

$$P_{av} = kTB,$$

where k is Boltzmann's constant, T is the temperature in degrees kelvin, and B is the equivalent noise bandwidth of the measurement system. Because maximum power is delivered into a resistance of equal value the delivered voltage will be $v_n/2$. The mean square value of v_n is

$$\text{Mean}[v_n^2] = 4\,kT\,R\,B.$$

The flat noise spectrum leads to a power spectral density

$$Sv_n(f) = 4\,kT\,R$$

per hertz.

10.3.3 Circuit Noise

Your circuit may have many noisy resistors, so you need to be able to relate the noise sources in the circuit to some measure of overall circuit performance. To do so, you could analyze the circuit for each individual noise source and put all the resulting noise powers into a noise figure or CNR formula. Adding up individual noise sources is a simple concept but a laborious procedure, however, and programming such a general method is difficult. Correlation matrices are not as simple a concept, but they are powerful and simple to use once understood. All in all, correlation matrices are a better way to compute noise performance.

Your next step is to relate the individual resistor noise sources to the correlation matrix. If you have a network of noisy resistors such as in Figure 10.2(a), you must first create an equivalent noiseless network that has noise sources external to the network, as shown in Figure 10.2(b).

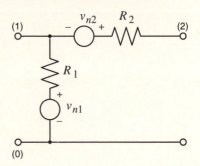

(a)

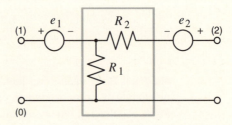

(b)

Figure 10.2 L-networks: (a) noise sources with resistors, and
(b) equivalent noise sources outside of network.

You can solve for the equivalent noise sources, e_1 and e_2, by open circuiting one port and measuring the noise voltage at the other port. Open-circuit voltage measurements will give the following expressions for e_1 and e_2:

$$e_1 = v_{n1},$$

$$e_2 = v_{n1} + v_{n2}.$$

Next you must compute the correlation matrix for the noise vector $\{e_1, e_2\}$.

In:
```
(cM = Outer[Cor, {e1,e2}, {e1,e2}]) //MatrixForm
```
Out:
```
Cor[e1, e1]    Cor[e1, e2]
Cor[e2, e1]    Cor[e2, e2]
```

A few function definitions will help you see how the correlation functions expand when multiple noise sources are considered. The following functions change the correlation of sums to the sum of correlations. When you substitute the internal noise sources, v_n, for the port noise sources, e, in **cM** the new **Cor** functions expand the result.

```
In:
  Cor[Plus[a_,b_],c_] := Cor[a,c] + Cor[b,c]
  Cor[c_,Plus[a_,b_]] := Cor[c,a] + Cor[c,b]

  (vM = cM /. {e1->vn1,e2->vn1+vn2})//MatrixForm
Out:
  Cor[vn1, vn1]    Cor[vn1, vn1] + Cor[vn1, vn2]

  Cor[vn1, vn1] + Cor[vn2, vn1]    Cor[vn1, vn1] +
      Cor[vn1, vn2] + Cor[vn2, vn1] + Cor[vn2, vn2]
```

Once you have made the substitution down to elementary sources, you can use the results of Sections 10.1 and 10.2 to simplify the results. The crosscorrelation of two different sources will be zero, and the autocorrelation of a single source will be the inverse Fourier transform of its spectral density. We can create a function, **SD**, that zeros the crosscorrelation terms and converts the autocorrelation terms to a power spectral density:

```
In:
  SD[Cor[a_,b_]] := 0
  SD[Plus[a_,b__]] := SD[a] + SD[Plus[b]]
  SD[Cor[a_,a_]] := SD[a]
  SetAttributes[SD,Listable]

  SD[vM]
Out:
  SD[vn1]             SD[vn1]
  SD[vn1]             SD[vn1] + SD[vn2]
```

Notice how all the terms involving **Cor[vn1,vn2]** are set to zero. **Cor[e1,e2]** will not always be zero. **e1** and **e2** may contain noise from the same source, as shown in the previous example, and so be correlated. Finally, you have to replace the spectral densities with the equivalent thermal noise values:

```
In:
  (czM = SD[vM] /. {SD[vn1]->4 kT R1,
                    SD[vn2]->4 kT R2}) //MatrixForm
Out:
  4 kT R1             4 kT R1
  4 kT R1             4 kT R1 + 4 kT R2
```

Remember that the preceding matrix comes directly from the spectral densities of the equivalent noise sources.

```
In:
  Cz = SD[MatrixForm[cM]]

Out:
  SD[Cor[e1, e1]    Cor[e1, e2]]
     Cor[e2, e1]    Cor[e2, e2]
```

`Cz` is the impedance correlation power spectral density matrix. An intriguing fact is that it is also 4 kT times the impedance matrix of our L-pad network. For passive reciprocal networks, Z- and Y-matrices have simple correlation matrices given by

$$C_z = 4\,kT\,\mathrm{Re}[Z]$$

$$C_y = 4\,kT\,\mathrm{Re}[Y]$$

Circuit component matrices and noise correlation matrices are similar in many ways — in the next section you will see how working with noise correlation matrices is similar to working with circuit component matrices. Once you have your correlation matrix, noise analysis becomes easy. Figures of merit such as noise figures are computed directly from correlation matrices, and you can compute the noise of combined devices by combining correlation matrices as discussed in the following section.

10.4 Noise Matrix Analysis

It may seem as though our study of correlation matrices is taking us away from analyzing circuits. However, this section will show you how the mathematically abstract correlation matrices are intimately connected with circuit analysis. You can directly relate the correlation matrices corresponding to various circuit matrix representations to circuit performance measures such as noise figure and CNR. Section 10.5 relates one type of correlation matrix to noise figure — an indicator of how much a circuit degrades your signal-to-noise ratio (SNR).

10.4.1 Noise Matrix Description

You can derive noise correlation matrices using a simple extension of circuit matrix analysis. The circuit of Figure 10.3 contains a two-port circuit and two external noise sources. Since series voltage noise sources are used, the circuit is described by an impedance matrix. The matrix equation describing Figure 10.3 is

$$\begin{bmatrix} V_1 \\ V_2 \end{bmatrix} = Z \begin{bmatrix} I_1 \\ I_2 \end{bmatrix} + \begin{bmatrix} e_1 \\ e_2 \end{bmatrix} .$$

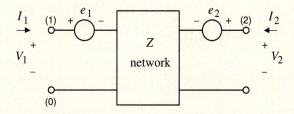

Figure 10.3 Impedance matrix with external noise sources.

The noise sources, given by e_1 and e_2, add a noise voltage vector to the standard circuit matrix analysis [Hillbrand76, Ambrozy82]. You should remember that e_1 and e_2 are in the frequency domain. Because the network is linear, matrix techniques can be used to combine several noisy networks or to convert a network between various representations. The noise sources do have one major complication: Since noise is random, you must perform a correlation or spectral density computation to convert the random signals to easily measured quantities.

For example, in Section 10.3, the spectral density computation for the correlation Z-matrix, **czM**, gave densities of **4kT R1**, **4kT(R1+R2)** and a cross-spectral density of **4KT R1**. You may be tempted to claim that individual noise sources have an RMS voltage of $\sqrt{4kTBR}$. Do not do it. Equating random variables to constants inevitably causes problems: Besides being fundamentally wrong, you lose all understanding of correlations.

In the next two sections we work with noise matrices and matrix mathematics and learn more about applying *Mathematica* to real problems.

10.4.2 Converting Correlation *Z*-Matrix to *Y*-Matrix

When you convert an impedance matrix circuit description to an admittance matrix description, all you really need to do is a matrix inversion. Once a noise vector is added, matrix conversions become a bit more involved but still follow the same process as regular circuit analysis. You begin with an impedance matrix description, as given in the following equation:

$$\begin{bmatrix} V_1 \\ V_2 \end{bmatrix} = Z \begin{bmatrix} I_1 \\ I_2 \end{bmatrix} + \begin{bmatrix} e_1 \\ e_2 \end{bmatrix},$$

$$\begin{bmatrix} V_1 - e_1 \\ V_2 - e_2 \end{bmatrix} = Z \begin{bmatrix} I_1 \\ I_2 \end{bmatrix},$$

$$Z^{-1} \begin{bmatrix} V_1 - e_1 \\ V_2 - e_2 \end{bmatrix} = \begin{bmatrix} I_1 \\ I_2 \end{bmatrix}.$$

Or, since $Y = Z^{-1}$,

$$\begin{bmatrix} I_1 \\ I_2 \end{bmatrix} = Y \begin{bmatrix} V_1 + e_1 \\ V_2 + e_2 \end{bmatrix},$$

$$\begin{bmatrix} I_1 \\ I_2 \end{bmatrix} = Y \begin{bmatrix} V_1 \\ V_2 \end{bmatrix} - Y \begin{bmatrix} e_1 \\ e_2 \end{bmatrix}.$$

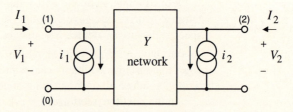

Figure 10.4 Admittance matrix with external noise sources.

Figure 10.4 shows the signal and noise variables for an admittance matrix. The noise current vector for the Y-matrix description is $\{i_1, i_2\} = -Y\{e_1, e_2\}$. You can compute the admittance correlation matrix **cy** by taking the spectral density and correlation of the outer product of the time-domain current noise vector, as shown below. The three steps to a frequency-domain correlation matrix (outer product, correlation, and spectral density) may seem like a lot of work. In reality, computing the frequency-domain correlation matrix is simple. The time dependence is shown explicitly in the equation below to help avoid confusion.

```
In:
  Cy == SD[MatrixForm[Outer[Cor,{i1[t],i2[t]},
                                {i1[t],i2[t]}]]]
```

```
Out:
  Cy == SD[Cor[i1[t], i1[t]]    Cor[i1[t], i2[t]]]
           Cor[i2[t], i1[t]]    Cor[i2[t], i2[t]]
```

You are using noise voltages and currents that are already in the frequency domain, so the analysis becomes simpler. If you look back at the functions **Correlation** and **PowerSpectrum**, you will see that the correlation was computed by using Fourier transforms. In fact, the **PowerSpectrum** can be computed directly by multiplying a frequency-domain signal by its complex conjugate. Once we are dealing with frequency-domain signals our correlation matrix becomes very simple: The correlation matrix in the frequency domain is just the outer product of the noise vector and its complex conjugate, as shown next. Note that the Fourier transform of each noise signal uses an infinite signal duration and effectively averages the noise signal at each frequency so the randomness is removed. In the time-domain view, the correlation function performs the averaging. In the following equation, **i1c[f]** is the conjugate of **i1[f]**:

```
In:
  Cy == (Outer[Times,{i1[f],i2[f]},
                     {i1c[f],i2c[f]}]//MatrixForm)
```

```
Out:
  Cy == i1[f] i1c[f]    i1[f] i2c[f]
        i1c[f] i2[f]    i2[f] i2c[f]
```

Now that you have the correlation matrix in the frequency domain, you can relate **Cy** to **Cz**. An outer product of a vector is a dot product of the vector and its transpose. When you substitute **-Y.{e1,e2}** for **{i1,i2}**, the transpose swaps the order of **{e1,e2}** and the **Y** matrix and you are left with the following *Mathematica* code for **Cy**:

```
e1c = Conjugate[e1];
e2c = Conjugate[e2];
Ytc = Transpose[Conjugate[Y]];

Cy == Y.Outer[Times,{e1,e2},{e1c,e2c}].Ytc;
Cy == Y.Cz.Ytc;
```

Because the outer product in the above equation is just **Cz**, you can see that **Cy** is related to both **Cz** and the circuit admittance matrix, **Y**

When you combine circuit matrices, the noise matrices must also be combined. For a series connection of two networks, you add the impedance matrices and you add the impedance correlation matrices. Similarly, a parallel connection of networks requires an addition of admittance matrices and correlation Y-matrices. Often, circuit and noise matrix combinations are not as simple as adding two matrices, but, in all cases, noise matrices are obtained by matrix mathematics and using the outer product to construct a correlation matrix.

Some noise matrices relate directly to measurement techniques and device characterization: S-parameter noise matrices relate to noise wave analysis and correlation $ABCD$-matrices relate to network noise figure.

10.4.3 The Correlation *ABCD*-Matrix

The correlation $ABCD$-matrix, C_{ABCD}, is important because it relates directly to the most commonly used network noise description: noise figure. Both noise figure and C_{ABCD} are defined using a series noise voltage and a shunt noise current at the input of a network (Figure 10.5).

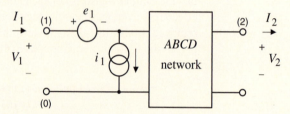

Figure 10.5 *ABCD*-matrix with noise sources.

As an example, we will convert **Cz** to **Ca**, where **Ca** is C_{ABCD} in *Mathematica* notation. When you calculate the **Ca** matrix, you can see how the diagonal terms are the power spectral densities of the individual sources and the off-diagonal terms are the cross-spectral densities. Again, **e1c[f]** is the conjugate of **e1[f]**.

```
In:
  Ca == (Outer[Times,{e1[f],i1[f]},
                    {e1c[f],i1c[f]}] //MatrixForm)

Out:
  Ca == e1[f] e1c[f]    e1[f] i1c[f]
        e1c[f] i1[f]    i1[f] i1c[f]
```

Your *ABCD*-matrix equation is:

$$\begin{bmatrix} V_1 \\ V_2 \end{bmatrix} = ABCD \begin{bmatrix} V_2 \\ -I_2 \end{bmatrix} + \begin{bmatrix} e_1 \\ i_1 \end{bmatrix} \ .$$

You need to solve for I_1 and eliminate I_1 from the equation for V_1 to convert the Z-matrix to the *ABCD*-matrix.

```
In:
   I1Soln = Solve[V2=={z21,z22}.{I1,I2} + e2,I1]
Out:
   {{I1 -> (-e2 + V2 - I2 z22)/z21}}
```

Substituting into our equation for V_1 gives

```
In:
   V1Soln =
        {z11,z12}.{I1,I2} + e1 /. I1Soln //Expand
Out:
   {e1 + I2 z12 - (e2 z11)/z21 + (V2 z11)/z21 -
        (I2 z11 z22)/z21}
```

When you collect the terms into the form for an *ABCD*-matrix you get

$$\begin{bmatrix} V_1 \\ I_1 \end{bmatrix} = \frac{1}{Z_{21}} \begin{bmatrix} Z_{11} & Z_{12}Z_{21}-Z_{11}Z_{22} \\ 1 & -Z_{22} \end{bmatrix} \begin{bmatrix} V_2 \\ -I_2 \end{bmatrix} + \begin{bmatrix} 1 & \frac{-Z_{11}}{Z_{21}} \\ 0 & \frac{-1}{Z_{21}} \end{bmatrix} \begin{bmatrix} e_1 \\ e_2 \end{bmatrix} .$$

The matrix

$$\begin{bmatrix} 1 & \frac{-Z_{11}}{Z_{21}} \\ 0 & \frac{-1}{Z_{21}} \end{bmatrix}$$

is the transformation matrix, *M*, for the *Z*-to-*ABCD* conversion. In the *Z*-to-*Y* conversion, *M* was simply equal to *Y*. The following *Mathematica* code gives the **Ca** matrix from the **CZ** and **Z** matrices:

```
Mtc = Transpose[Conjugate[M]];

Ca == M.CZ.Mtc
```

If you use the resistive L-pad in Section 10.3 as an example, you will have the following **M** and **Ca** matrices. **m** is the temporary variable for the **M** matrix in the following equation:

```
In:
  m = {{1,-R1/R1},{0,-1/R1}};

  Ca == Simplify[4 kT M.{{R1,R1},{R1,R1+R2}}.
                  Transpose[Conjugate[M]] /.
                       M->m] //MatrixForm

Out:
  4 kT R2                    (4 kT R2)/R1

                                               2
  (4 kT R2)/R1               (4 kT (R1 + R2))/R1
```

You cannot just use e_1 from the Z-matrix description and i_1 from the Y-matrix description. The movement of noise sources through the network changes in correlations between the sources and their powers.

Noise correlation matrices do three things for us: The matrices contain a complete component noise description, the matrices can be related to any system noise specification, and the matrices have all the information we need for imbedding our component noise performance into a larger system of components.

10.5 Noise Figure

Each area of engineering has its favorite measure of circuit performance. Typically, high-frequency component engineers use gain and noise figure. Communication system designers may use CNR for analog systems and BER for digital systems. In the following sections, we will examine noise figure in detail. We begin by defining noise figure, and then we use a general circuit to relate noise figure to correlation matrices. The resulting relationship between noise figure and the C_{ABCD} correlation matrix is very simple, but the algebra of the next few sections would be tedious to do by hand. By using **Solve**, **D**, and **Simplify**, we can get *Mathematica* to make our work much simpler.

10.5.1 Signal-to-Noise Degradation

The following noise figure equation uses six network parameters: F_{min}, R_n, Y_{opt}, and Y_s. Because Y_{opt} and Y_s are complex, they count as four parameters. The dependence of noise figure on source impedance is often held separate from the other four noise parameters. The interaction between the optimum source admittance, Y_{opt}, and the source admittance in use, Y_s, determines how far the noise figure is above the minimum noise figure, NF_{min}. The following equation describes noise figure as a power ratio, F [Ambrozy82]. Typically, F refers to noise figures as a power ratio and NF refers to noise figures in dB.

$$F = F_{min} + \frac{R_n}{G_s}\left|Y_s - Y_{opt}\right|^2$$

G_s is the source conductance or the real part of Y_s. When noise figure is given in dB, the following equation holds:

$$NF = 10\,Log_{10}(F).$$

Noise figure can also be defined as a degradation in signal-to-noise ratio. After some manipulation, you can derive that F is just the total noise referred to the device input divided by the noise from the source. The dependence on source noise also means noise figure implies a source temperature since the source noise is almost always from a 50 or 75 Ω resistor. The IEEE reference temperature is 290 K [Miller67]. In the following manipulation, S_{in} is the input signal power, N_{in} is the input noise power, G_p is the network power gain, and N_{net} is the network output power:

$$\begin{aligned}
F &= (S/N)_{in}/(S/N)_{out} \\
&= ((S_{in}G_p)/(N_{in}G_p))/((S_{in}G_p)/(N_{in}G_p + N_{net})) \\
&= (N_{in}G_p + N_{net})/(N_{in}G_p) \\
&= (N_{in} + N_{net}/G_p)/N_{in}
\end{aligned}$$

10.5.2 Noise Figure and Noise Sources

You can compute the noise figure by dividing the total input noise power by the source noise power as given below (see Figure 10.6). Note that Y_{in} divides out of the final equation.

$$F = \left(|V_{in}|^2_{total} \right) \Big/ \left(|V_{in}|^2_{source} \right)$$

$$= \frac{\left| \dfrac{-i_s - i_n}{Y_s + Y_{in}} - \dfrac{e_n Y_s}{Y_s + Y_{in}} \right|^2}{\left| \dfrac{-i_s}{Y_s + Y_{in}} \right|^2}$$

$$= \frac{\left| i_s + i_n + e_n Y_s \right|^2}{\left| i_s \right|^2}$$

$$= \frac{|i_s|^2 + |i_n|^2 + |e_n Y_s|^2 + i_n e_n{}^* Y_s{}^* + i_n{}^* e_n Y_s}{|i_s|^2}$$

$$= 1 + \frac{|i_n|^2 + |e_n Y_s|^2 + i_n e_n{}^* Y_s{}^* + i_n{}^* e_n Y_s}{|i_s|^2}$$

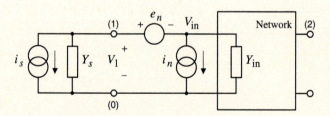

Figure 10.6 Network with noise sources referred to input.

You can rewrite the preceding equation in terms of the C_{ABCD}-matrix as follows, with the substitution that $|i_s|^2 = 4\,kT\,G_s$:

$$F = 1 + \frac{C_{a22} + C_{a11}|Y_s|^2 + C_{a21}Y_s{}^* + C_{a12}Y_s}{4kTG_s} \quad .$$

The C_{aij} terms are the four components of the C_{ABCD}-matrix. The four real numbers representing network noise at any frequency may be given in many ways. One way is: e_n^2, i_n^2, $\text{Re}[e_n i_n{}^*]$, and $\text{Im}[e_n i_n{}^*]$. Only the off-diagonal terms of the correlation matrix can be complex. Another way of writing the four noise parameters is: F_{min}, R_n, G_{opt}, B_{opt}. Once you compute the C_{ABCD}-matrix, you know the network noise figure for any source admittance. The only relation left to establish is between the preceding equation for F and the following more common form for F. You can use *Mathematica* to do most of the work for you.

$$F = F_{min} + \frac{R_n}{G_s}\left|Y_s - Y_{opt}\right|^2$$

10.5.3 Noise Figure Relationships

You can relate the *ABCD* correlation matrix to F_{min}, R_n, G_{opt} and B_{opt}. Once you minimize F with respect to G_s and B_s, you get equations for F_{min}, G_{opt}, and B_{opt}. In the following equation, we use $Y_s = G_s + I B_s$, $C_{a12} = C_{a21}{}^*$, and $C_{a12} = C_{a12r} + I C_{a12i}$:

```
In:
  FA = 1 + (Ca22 + Ca11 (Gs^2+Bs^2) +
                (Ca12r - I Ca12i) (Gs-I Bs) +
                (Ca12r + I Ca12i) (Gs+I Bs))/(4 kT Gs);

  BSoln = Solve[D[FA, Bs] == 0, Bs]

Out:
  {{Bs -> Ca12i/Ca11}}
```

So B_{opt} is $\text{Im}[C_{a12}]/C_{a11}$. For G_{opt}, you must solve the following equation:

```
In:
  GSoln = Solve[D[Simplify[FA /. BSoln][[1]], Gs]
                    == 0, Gs]

Out:
                           2
  {{Gs -> Sqrt[-(Ca12i /Ca11) + Ca22]/Sqrt[Ca11]},

                           2
  {Gs -> -(Sqrt[-(Ca12i /Ca11) + Ca22]/Sqrt[Ca11])}}
```

You will use the positive (first) solution for G_s. You can find F_{min} by substituting **GSoln** (G_{opt}) and **BSoln** (B_{opt}) into **FA**:

In:
    FminSoln = Simplify[FA /.
                              {GSoln[[1,1]],BSoln[[1,1]]}]

Out:

```
                               2
   1 + (Sqrt[Ca11] ((-2 Ca12i )/Ca11 + 2 Ca22 +

                            2
          (2 Ca12r Sqrt[-(Ca12i /Ca11) + Ca22])/

     Sqrt[Ca11])))/

                         2
      (4 Sqrt[-(Ca12i /Ca11) + Ca22] kT)
```

So our noise parameters become:

In:
    FminA =
      1 + (Sqrt[-Ca12i^2+Ca22 Ca11] + Ca12r)/(2 kT);
    GoptA = Sqrt[Ca22/Ca11 - (Ca12i/Ca11)^2];
    BoptA = Ca12i/Ca11;

The last parameter is R_n. If you substitute the parameters above into $R_n = (F - F_{min}) \, G_s \, / \, |Y_s - Y_{opt}|^2$, you get the following equation for R_n:

In:
    Simplify[Gs (FA - Fmin)/
               ((Gs - Gopt)^2 + (Bs - Bopt)^2) /.
      {Gopt -> Sqrt[Ca22/Ca11 - (Ca12i/Ca11)^2],
      Bopt -> Ca12i/Ca11,
      Fmin -> 1 + (Sqrt[-Ca12i^2 + Ca22 Ca11] +
                          Ca12r)/(2 kT)}]

Out:

```
      2
   (Bs  Ca11 - 2 Bs Ca12i + Ca22 -

                    2                           2
       2 Sqrt[-Ca12i  + Ca11 Ca22] Gs + Ca11 Gs )/

                              2
      (4 ((Bs - Ca12i/Ca11)  +
                          2              2     2
          (-Sqrt[(-Ca12i  + Ca11 Ca22)/Ca11 ] + Gs) ) kT)
```

Expanding the denominator of the preceding equation allows a few more terms to cancel and you will find that R_n is given by the simple equation

```
In:
   RnA = Ca11/(4 kT);
```

From the L-pad example in Section 10.3, you get the following set of noise parameters:

```
In:
   LPadSoln = {Ca11->4 kT R2,Ca12r->4 kT R2/R1,
               Ca12i->0,Ca22->4 kT (R1+R2)/(R1^2)};
   {FminA,RnA,GoptA,BoptA} /. LPadSoln //Simplify
Out:
                                2               2
   {1 + (2 R2)/R1 + (2 Sqrt[(kT  R2 (R1 + R2))/R1 ])/kT, R2,

                         2
      Sqrt[(R1 + R2)/(R1  R2)], 0}
```

As a numerical example with an **R1** of 150 Ω and an **R2** of 20 Ω, you get:

```
In:
   % /. {R1->150.0,R2->20.0, kT->4 10^-21}
Out:
   {2.04413, 20., 0.0194365, 0}
```

You should notice how R_n is **R2** and how B_{opt} is zero, because C_{a12} has no imaginary part. C_{a12} is the crosscorrelation of e_n and i_n. So reactive source admittances are needed only when e_n and i_n are correlated and the correlation has an imaginary part.

10.6 Noise Solutions with *Nodal*

The *Nodal* package has functions for both circuit matrix analysis (such as **YParameters**) and circuit noise analysis (such as **CyParameters**). Preprogrammed packages save you a lot of time and effort. The following example uses the **NoiseParameters** function to analyze the resistive L-pad of Figure 10.2. Of course, all the results are the same as those you computed previously.

The **NoiseParameters** function returns a list of data as a set of rules so you do not get confused over which numbers belong to what variable:

```
In:
   Needs["Nodal`","Nodal2.m"];
```

```
In:
   NoiseParameters[
      NodalNetwork[Resistor[{1,2}, 20.0],
                   Resistor[{1,0}, 150.0]]]

Out:
   Nodal`Fmin -> 2.04413
   NFmin -> 3.10508
   Nodal`Rn -> 20.
   Yopt -> 0.0194365
   Gammaopt -> 0.0142886
```

One important aspect of noise figure is that a plot of noise figure versus source impedance approximates a parabolic surface (when the noise figure is close to the optimum). The noise figure equation in Section 10.5.1 shows that variations in B_s create a parabola whereas variations in G_s only approximate a parabola. The pointedness of the parabola is proportional to R_n. **Plot3D** and **NoiseFigure** are used in the next example to plot the noise figure surface for an L-pad. The plot varies the source impedance rather than source admittance, but the fundamentals are the same. The minimum useful noise figure occurs at a source resistance of 51.5 Ω and no reactance.

```
In:
   Plot3D[
      DBP[NoiseFigure[x + I y,
                      NodalNetwork[
                         Resistor[{1,2}, 20.0],
                         Resistor[{1,0}, 150.0]]]],
      {x,10,150},
      {y,-100,100},
      AxesLabel->{"Rs","Xs","NFdB"},
      Boxed->False];

Out:
```

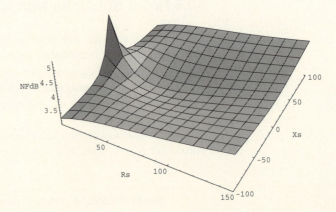

10.7 Summary

In this chapter, you have seen how

- to develop numerical functions to filter, correlate, and find the power spectrum of noise signals
- to characterize noise signals with statistical and frequency-domain methods
- to analyze electrical network noise using noise correlation matrices
- to analyze circuits using the high-level functions in *Nodal*.

Each section of this chapter used a different aspect of *Mathematica*. Section 10.1 used the standard statistical and plotting packages included with *Mathematica*. Pure functions were used in Section 10.1 for efficiently filtering noise signals in the time domain. In Section 10.2, you developed numerical functions for correlation and spectral density computation. Section 10.3 contained functions for symbolically manipulating noise signals. Both Sections 10.4 and 10.5 used *Mathematica* as an equation solver. Finally, Section 10.5 used *Nodal* to compute and plot circuit noise characteristics.

10.8 Exercises

10.1 Analyze the cross-spectral density and correlation of `white2` and `brown` waveforms.

10.2 Show how the correlation Y-matrices add when two networks are paralleled.

10.3 Given a Z-matrix and a C_Z-matrix, find the total output noise voltage when the input is shorted.

10.4 Use the nonlinear filter $v_{Out} = v_{In}^2$ on white noise. Plot the histogram and the power spectral density of v_{Out}. What has changed and why?

10.5 Determine the rule for cascading C_{ABCD} matrices.

10.6 Find the C_Y matrix of a T-pad.

10.7 Use *Nodal* to compute the noise parameters of a series 20 Ω resistor and shunt 20 pF capacitor at 100 MHz. Why is `Yopt` complex?

10.8 Use *Nodal* to evaluate the noise parameters of a bipolar transistor. Use a manufacturer's data book to find the device resistances, capacitances, and typical bias current. Compare the noise parameters given by *Nodal* with those of the data book.

10.9 References

[Ambrozy82] Ambrozy, A., *Electronic Noise*, McGraw-Hill, New York, 1982.

[Carlson81] Carlson, A.B., *Communication Systems*, McGraw-Hill, London, 1981.

[D'Amico86] D'Amico, A., Mazzetti, P., *Noise in Physical Systems and 1/f Noise*, Elsevier, Amsterdam, 1986.

[Gardner78] Gardner, M., "White and brown music, fractal curves and one-over-*f* fluctuations," *Scientific American*, pp. 17–32, April 1978.

[Haus59] Haus, H.A., Adler, R.B., *Circuit Theory of Linear Noisy Networks*, John Wiley & Sons, New York, 1959.

[Hillbrand76] Hillbrand, H., Russer, P. H., "An Efficient Method for Computer Aided Noise Analysis of Linear Amplifier Networks," *IEEE Trans. CAS*, pp. 235–238, April 1976.

[Johnson71] Johnson, J.B., Electronic Noise: The First Two Decades, *IEEE Spectrum*, 8:42–46, 1971.

[Keshner82] Keshner, M. S., "1/*f* Noise," *Proc. IEEE*, 70(3): 212–218, 1982.

[Miller67] Miller, C. K. S., Daywitt, W. C., Arthur, M. G., "Noise Standards, Measurements, and Receiver Noise Definitions," *Proc. IEEE*, 55:865–877, 1967.

[National76] Bohn, D., *Audio Handbook*, National Semiconductor, Santa Clara, California, 1976.

[Oppenheim75] Oppenheim, A.V., Shafer, R.W., *Digital Signal Processing*, Prentice-Hall, Englewood Cliffs, New Jersey, 1975.

[Pickholtz82] Pickholtz, R. L., Schilling, D. L., Milstein, L. B., "Theory of Spread-Spectrum Communications – A Tutorial," *IEEE Trans. Comm.*, COM-30:855–884, 1982.

[Shannon48] Shannon, C.E., "A Mathematical Theory of Communications," *BSTJ*, 27:379–423, and 27:623–656, 1948.

[Wolfram91] Wolfram, S., *Mathematica — A System for Doing Mathematics by Computer*, Addison-Wesley, Redwood City, California, 1991.

Appendix

In this Appendix, we have collected together useful commands and functions from *Mathematica*, *Nodal*, and this book. We do not intend the collection to be a replacement for the detailed information supplied with *Mathematica* and the full version of *Nodal*, but we feel that it is often useful to have the most commonly used commands collated in one handy reference.

A.1 *Mathematica* Functions

A.1.1 Syntax

Mathematica treats any text between (* and *) as a comment (that is, *Mathematica* ignores the intervening characters).

Commands and functions have names beginning with upper-case characters. Arguments are separated by commas and enclosed in square brackets: `Command[`*arg1*`,`*arg2*`]`.

Mathematica's syntax has a lot in common with many programming languages. For example, you can insert any number of spaces between symbols but not within a symbol: `x y` means `x` times `y`, whereas `xy` is a symbol with the name `xy`. Having a space imply multiplication is rare in programming languages but common in mathematics. In *Mathematica*, as in most other programming languages, you can also use the * symbol to denote multiplication. For general syntax information, we strongly recommend that you browse through the "Basic Objects" section in Stephen Wolfram's book *Mathematica — a System for Doing Mathematics by Computer*.

Mathematica contains a large number of special forms for functions. For example, `a=3` is the special form of the function `Set`, and this operation can be written `Set[a, 3]`. Also, `=` is the infix form of `Set`. Many of the special forms are normal mathematical and/or computing notations (`x^y` is equivalent to `Power[x, y]` or x^y), but others, such as `//@`, are indigenous to *Mathematica* and you may find them confusing at first. Remember that the special forms are merely shorthand notation and that you can always use their full forms if you wish. You can obtain help on both special and full forms by preceding them by a `?`:

```
In:
  ?//@

Out:
  MapAll[f, expr] or f //@ expr applies f to every
     subexpression in expr.
```

A.1.2 Numbers

Numbers can be integers (**1**), or rational numbers (**4/5**), approximate real numbers (**1.**), or complex numbers (**1. + I 5**). Note that *Mathematica* sometimes cannot simplify expressions containing real numbers because it regards them as being imprecise.

Mathematica uses the **^** symbol for exponentiation: **10^7** is equal to ten million. To raise *e* to a power, you can use **Exp**[*power*] or **E**^*power*.

Do not use the method, often adopted in other computer languages, where **e** or **f** (for example) indicate the beginning of the exponentiation value. In *Mathematica*, **1.2e4** means "1.2 times the variable called e4," not, as in BASIC or FORTRAN, "1.2 times ten-to-the-fourth-power."

To assign a value to a variable, use *variableName* = *value,* and to deassign a value from a variable, use *variableName* =. , or **Clear**[*variableName*] (that is, a full stop following an equals sign).

In *Mathematica*, you can use the **%** symbol to refer to the last answer. Using **%** saves you from having to name results and is convenient when you wish to refer only once to a result that has been calculated by the previous operation.

Rationalize[*x*] converts a floating point number, *x*, into the nearest rational number where possible. **Rationalize**[*x, dx*] gives the nearest rational number to a precision *dx*; if *dx* = 0, the accuracy of *x* serves as a guide for the acceptable accuracy of the result. **Rationalize** is useful for converting decimal numbers into transformer turns ratios.

The numerator and denominator of a fraction can be extracted using the **Numerator** and **Denominator** functions.

N[*argument*] forces *Mathematica* to give the real number form of *argument*.

A.1.3 Lists

A list is a collection of items (for example, numbers or other lists) and is enclosed in curly brackets: {*item1*, *item2*, *item3*} is a one-dimensional list and {{*a1*,*b1*},{*a2*,*b2*},{*a3*,*b3*}} is a two-dimensional list.

Range[*max*] generates a list from 1 to *max* in steps of 1. **Range**[*min, max,delta*] will generate a list from *min* to *max* in steps of *delta*.

Table[*expression*,{*var*,*min*,*max*,*delta*}] generates a list of expressions evaluated from *min* to *max* in steps of *delta*.

A.1.4 Manipulating Lists

Given a list **L** with **n** members, you can identify the *i*th member by **L**[[*i*]], where **1** ≤ *i* ≤ **n**.

Two or more lists can be joined together (concatenated) using **Join**[*list1, list2,...*]. A list can be partitioned into smaller lists of length **len** using **Partition**[*list,* **len**].

Transpose[*list*] reverses the indices of the elements in *list*. For example, {{*a1,b1*},{*a2,b2*},{*a3,b3*}} becomes {{*a1,a2,a3*},{*b1,b2,b3*}}.

Reverse[*list*] reverses the order of the elements in *list*.

Take[*list, n*] returns the first *n* elements of *list*. If *n* is negative, then the last *n* elements are taken.

Drop[*list, n*] returns *list* with the first *n* elements removed. If *n* is negative, then the last *n* elements are removed.

Join[*list1, list2,...*] concatenates its arguments into one list.

A.1.5 Basic Arithmetic Operations

The operators **+** and **–** have their usual meaning. *Mathematica* uses a space or the ***** character as the multiplication operator. Division is accomplished using the slash (**/**) character.

A.1.6 Complex Numbers

Use **I** as $\sqrt{-1}$: A complex number is represented in the form **x + I y**. For a complex number z, its real and imaginary parts can be extracted using **Re**[*z*] and **Im**[*z*].

The absolute value ($|z|$), argument ($|z|e^{i\phi}$) in radians, and conjugate ($z*$) can be found using the functions **Abs**[*z*], **Arg**[*z*], and **Conjugate**[*z*], respectively.

A.1.7 Rule Symbol and Replacement

The character **->** (typed as a hyphen followed by a right angle bracket) is *Mathematica*'s rule symbol: **x->3** means make **x** take the value of **3** only in the command or function in which the rule is being used. (If you wish to assign **x** the value **3**, then use **x=3**.) Both right and left sides of the rule symbol can be symbolic. Lists of rules are enclosed in **{}**.

The infix form of the **ReplaceAll** operator, **/.**, replaces symbols in an expression on its left by values described by rules (such as *symbol->value*) on its right: *expression* **/.** *rules*. For example, **Sin[x] /. x->3** would return **Sin[3]** and leave **x** as a variable in all other expressions.

A.1.8 Manipulating Expressions

You may often find that the **Simplify**[*expression*] function clarifies complicated expressions. Other useful *Mathematica* commands that alter the form of an answer are **Expand**, **ExpandAll**, **Factor**, **Together**, **Apart**, and **Cancel**.

A.1.9 Prefix, Infix, and Postfix Forms of Operators

As an alternative to the standard format for operator and argument, operators can be used in prefix, infix, and postfix forms:

standard: **Operator** [*argument*]

prefix: **Operator @** *argument*

infix: *arg1* **~Operator~** *arg2*

postfix: *argument* **//** **Operator**

A.1.10 Matrix Multiplication

Matrices can be multiplied together using the **Dot** [*arg1, arg2*] operator which can be contracted to its infix form, **.**, to mimic normal mathematical notation: *arg1* **.** *arg2*

A.1.11 User-Defined Functions

User-defined functions have the general form **f[x_]:=***function* or, for specific types of argument **x_** becomes **x_***type* where *type* is one of *Mathematica*'s recognized types (the *type* is actually the head of **x**).

A.1.12 Reading Data from ASCII Files

Use **ReadList** [*"filename", type*] to read in data of a given type from the file called *filename*.

A.1.13 Fitting Data to Functions

Fit [*data, function, var*] fits a function to data as a dependent of the variable *var*.

A.1.14 Manipulating Equations

The **Solve** [*eqnList, varList*] function requires two arguments: a list of equations to solve and a list of the variables in terms of which a solution is required. Equations are defined using the **==**, **Equal**, operator. A list of variables to be eliminated is an optional third argument.

    FindRoot [*expression, {var, startValue}*] will find (numerically) the value of *var* (starting at *var=startValue*) at which the *expression* is zero.

    FindMinimum [*function, {var, startValue}*] will find (numerically) the value of *var* (starting at *var=startValue*) at which the *function* is at a local minimum near *startValue*.

A.1.15 Calculus

D[_function_, _variable_**]** gives the partial derivative of _function_ with respect to _variable_.

Integrate[_expr_,_var_**]** gives the indefinite integral of _expr_ with respect to _var_.

Integrate[_expr_, {_var_, _min_, _max_}**]** gives the definite integral of _expr_ with respect to _var_ over the range from _min_ to _max_.

A.1.16 Anonymous Functions

Although you can declare user-defined functions using the format **fName[**_arg__**]:=**_functionBody_, _Mathematica_ allows a shorter version that does not employ explicit names for arguments or for the function. You may find this shorter version useful when a user-defined function is helpful but will not be used elsewhere. The shorter form uses the **Function** command (or its postfix equivalent **&**) with argument positions indicated by **#** for single arguments, **##** for arguments that are lists, or **#n** where it is the **n**th argument in a list that is to be used by the function. This is all very complicated until you try it — when, we hope, you will suddenly find it much simpler!

For example, if you require an argument, **arg**, to be squared, you might declare it as **(#^2)& arg**, which means "take a function **#^2** and substitute the argument for the symbol **#**."

A function to calculate a triangle's hypotenuse requires two arguments. You can specify which argument slots into which part of your function by using **#n**. Note that the arguments for the function are declared in square brackets, just as they would be for a named function. **#1** refers to the first argument (**a**) and **#2** to the second (**b**):

```
hyp[a_,b_]:=Sqrt[(#1^2 + #2^2)&   [a,b]]
```

A.1.17 Random Number Generation

Random[] returns a random real number in the range 0 to 1.

Random[_type_, _max_**]** and **Random[**_type_, {_min_,_max_}**]** return random numbers between 0 and _max_ and between _min_ and _max_, respectively, of the specified _type_. _type_ can be one of **Real**, **Complex**, or **Integer**. An optional third argument to **Random** specifies the precision for the result.

Use **SeedRandom[]** to reseed the random-number generation; an optional argument can be used to specify the seed if you require a repeatable random sequence.

A.1.18 Laplace Transform

After loading the standard Laplace transform package via
`<<Calculus`LaplaceTransform``, the function
`LaplaceTransform[`*expr*`,`*t*`,`*s*`]` will compute the Laplace transform of *expr*
as a function of *t*; the inverse operation is performed by
`InverseLaplaceTransform[`*expr*`,`*s*`,`*t*`]`.

A.1.19 Manipulating Polynomials

`Factor[`*poly*`]` factors a polynomial *poly* that contains integer or rational coefficients. `Factor` does not work with real number coefficients. Where the coefficients of *poly* are complex, use the option `GaussianIntegers->True`.

`Together[`*expr*`]` collates the terms in *expr* over a common denominator and
cancels common factors.

`Apart[`*expr*`]` reformats *expr* as a sum of terms with simple denominators.

`Expand[`*expr*`]` expands products and positive integer powers in *expr*.
`ExpandAll` expands out all products and powers; the functions
`ExpandNumerator` and `ExpandDenominator` manipulate specific parts of
an expression.

A.2 *Nodal* Components, Functions, Utilities, and Constants

This section is intended as a guide to features in *Nodal* — not as a replacement
for the manual, which can contain more up-to-date information.

A.2.1 Components

This section contains an alphabetical list of all the components in *Nodal* (version
2.0). Where a component has an option, we have described it. All values are in SI
units (for example, ohms, farads, henries, volts, amperes, mhos). All components
except `SubNetwork` can be drawn using the `Draw[`*ComponentName*`[]]`
command. You may find that drawing provides a useful memory-jogger for both
connections and the internal elements of the more complicated components.

`Admittance[{`*node1*`,`*node2*`},` *value*`,` *options*`]`

Specifies an admittance. For frequency-dependent values, use `f` as the variable
for frequency.
Options: `Temperature->t` (`t` in Kelvin).

BJT[{*base*,*emitter*,*collector*}, *rb*, *Re*, *Ce*, *Cbc*, *beta*, *ftHz*]

Bipolar junction transistor using six-term bridged-T model.

Capacitor[{*node1*,*node2*}, *value*]

Specifies simple (ideal) capacitor. *See* **CapacitorModel** for a more elaborate model.

CapacitorModel[{*node1*,*node2*}, *value*, *options*]

Uses a series combination of C, R, L to synthesize a nonideal capacitor. *Options:* Temperature can be set using **Temperature->t** with **t** in kelvin. The resistance and inductance can be specified using one of two methods:
R->resistorValue, L->inductorValue
or, by analogy with a lossy tuned circuit,
Q->qValue, fo->resonantFrequency
For example,

```
CapacitorModel[{3,2}, 10 uF]
CapacitorModel[{3,2}, 10 uF, R->0.4, L->1 nH]
CapacitorModel[{3,2}, 10 uF, Q->200, fo->16.7 MHz]
CapacitorModel[{3,2}, 10 uF, Q->100, fo->170 kHz,
                                    Temperature->200]
```

Capacitor Types: *See* **ParallelPlate**, **Gap**, and **Interdigital** in Section A.2.3.

CCCS[{*node1*,*node2*,*node3*,*node4*}, *gain*, *fc*]

Specifies current-controlled current source. Current flowing from *node4* to *node3* is the product of *gain* and the current flow from *node1* to *node2*. *gain* defines the multiplication factor. *fc* is the corner frequency in Hz and is optional; above *fo*, the gain decreases by 20dB/decade. An output resistance of 1 mΩ is used for *Y* analysis.

CCVS[{*node1*,*node2*,*node3*,*node4*}, *tr*, *fc*]

Specifies a current-controlled voltage source. Voltage across *node4* (+) and *node3*(−) is the product of *tr* and the current flowing from *node1* to *node2*. *fc* is the corner frequency in Hz and is optional; above *fo*, the *tr*-current product decreases by 20 dB/decade. An output resistance of 1 mΩ is used for *Y* analysis.

Circulator[{*node1*,*node2*,*node3*}, *loss*, *isolation*, *options*]

Circulation direction is *node1* to *node2* to *node3*. *loss* and *isolation* are specified in dB.

Options: The impedance of the device can be defined using **Zo->value**; **Zo->50** is the default.

Conductor Types: *See* **Ribbon** and **Wire** in Section A.2.3.

CoupledInductors[{node1**,**node2**,**node3**,**node4**}, **L1**, **L2**, **k]**

Simple transformer topology. First inductor of value *L1* from *node1* to *node2*, second inductor of value *L2* from *node4* to *node3*. *k* (<1) is the coupling coefficient: $M = k\sqrt{L1L2}$.

CoupledInductors3[{node1**, **node2**, **node3**, **node4**, **node5**, **node6**},
    L1, **L2**, **L3**, **k12**, **k23**, **k13]**

Simple three-winding transformer topology. First inductor of value *L1* from *node1* to *node2*, second inductor of value *L2* from *node4* to *node3*, and third inductor of value *L3* from *node6* to *node5*. k_{ij} (k < 1, i ≠ j) is the coupling coefficient between the *i*th and *j*th inductor, such that: $M_{ij} = k_{ij}\sqrt{L_iL_j}$.

CurrentNoise[{node1**,**node2**}, **iNoiseRMS**, **options]**

Defines a current noise source, current flowing from *node2* to *node1*. *iNoiseRMS* is the rms noise current per Hz.
Options: **Temperature->t**; t=300 kelvin by default.

CurrentSource[{node1**,**node2**}, **current]**

Specifies a current source flowing from *node2* to *node1*. Where *current* is a function of frequency, use **f** as the frequency variable.

Diode[{node1**,**node2**}, **Rd**, **Cd**, **Rs**, **options]**

Defines a diode with dynamic resistance *Rd* in series with *Rs*, the parasitic and contact resistance. The junction capacitance *Cd* is in parallel with *Rd*.
Options: Diode temperature can be given by specifying **Temperature->t**; t=300 kelvin by default.

FETs: *See* JFET, MESFET, or MOSFET.

Gyrator[{node1**,**node2**,**node3**,**node4**}, **fcg**, **bcg]**

Defines a gyrator with forward current gain *fcg* and backward current gain *bcg*. Gyrator elements between (a) *node1* and *node2* and (b) *node3* and *node4*.

Impedance[{*node1*,*node2***}, *z*, *options***]**

Specifies a generic impedance connected between *node1* and *node2* with value *z*. If *z* is frequency dependent, use **f** as the frequency variable.
Options: Temperature given by **Temperature->t**; **t** = 300 kelvin by default.

Inductor[{*node1*,*node2***}, *h***]**

Simple single inductor of value *h* connected between *node1* and *node2*. *See* **InductorModel** for a more generalized model.

InductorModel[{*node1*,*node2***}, *h*, *options***]**

Pure inductance of value *h* with series resistance in parallel with a capacitance.
Options: Temperature can be set using **Temperature->t**; **t** = 300 kelvin is the default. The value of the series resistance and capacitance can be specified using one of two methods:
R->resValue, C->capValue
or, by analogy with a tuned circuit,
Q->qValue, fo->resonantFrequency
For example,

```
        InductorModel[{1,2}, 1 uH]
        InductorModel[{1,2}, 2.3 mH, R->10, C->10 pF]
        InductorModel[{1,2}, 2.3 mH, R->10, C->10 pF,
                    Temperature->273]
        InductorModel[{2,3}, 4.3 mH, Q->100, fo->1.5 MHz]
```

Inductor Types: *See* the components **CoupledInductors** and **CoupledInductors3** as well as the utilities **Ribbon**, **Wire**, **ParallelWire**, **SingleLayerSolenoid**, **Solenoid**, **Toroid**, **Spiral**, **ViaHole**.

Isolator[{*node1*,*node2***}, *lossDB*, *isolationDB*, *options***]**

Defines an isolator with loss and isolation (in dB) specified.
Options: The impedance of the isolator can be set using **Zo->zValue**; **zValue->50** Ω is the default.

JFET[{*gate*,*source*,*drain***}, *Rds*, *Cgs*, *Cgd*, *Cds*, *gm*, *options***]**

Defines a five-term FET model.
Options: Pucel/van der Ziel noise parameters can be assigned using **{P->pValue, R->rValue, C->cValue}**. By default, **pValue**=0.67, **rValue**=0.12, **cValue**=0.4.

MESFET[{_gate_**,**_source_**,**_drain_**}, **_Rgs_**, **_Rds_**, **_Cgs_**, **_Cgd_**, **_gm_**, **_Tau_**, **_options_**]**

Defines a seven-term FET model. _Tau_ is the delay in _gm_.
Options: Pucel/van der Ziel noise parameters can be assigned using
{P->pValue, R->rValue, C->cValue}. By default, **pValue**=1.2,
rValue=0.4, **cValue**=0.9. The temperature of _Rgs_ can be set by using the
option **TemperatureRgs->tValue**.

MOSFET[{_gate_**,**_source_**,**_drain_**,**_bulk_**}, **_Rds_**, **_Cgs_**, **_Cgd_**, **_Cds_**, **_Cbs_**, **_gm_**, **_options_**]**

Defines a six-term FET. High frequency applications can require the addition of
an external gate resistance.
Options: Pucel/van der Ziel noise parameters can be assigned using
{P->pValue, R->rValue, C->cValue}. By default, **pValue**=0.67,
rValue=0.12, **cValue**=0.4.

OpAmp[{_nonInvNode_**, **_invNode_**, **_outNode_**}]**

Specifies an ideal operational amplifier. Gain and output resistance are 10^6 and
10^{-3} Ω, respectively, for matrix analysis and infinity and zero for voltage and current analysis, respectively. _See_ **OpAmpModel** for a more generalized model.

OpAmpModel[{_nonInvNode_**, **_invNode_**, **_outNode_**}, **_Rin_**, **_Rout_**, **_gain_**, **_fbk_**,
        options]**

Generalized operational amplifier model. Input and output impedance is specified
by _Rin_ and _Rout_. _fbk_ is the frequency at which a 20 dB/decade decrease in _gain_
begins.
Options: You can specify the transfer function by giving it as the argument to the
rule assignment **TransferFunction->**; the default transfer function is
TransferFunction -> 1/(1+ I*f/fbk). In the **TransferFunction**
definition, **f** must be used for frequency and **fbk** for the value of the gain-
frequency knee; _Nodal_ will pick up the value of **fbk** from the assignment in
OpAmpModel's argument list. For example,

```
OpAmpModel[{3,2,6}, 1 MOhm, 100, 3*10^5, 25]
OpAmpModel[{3,2,6}, 1 MOhm, 100, 3*10^5, 25,
                TransferFunction->1/(1+f/(2 fbk))]
```

Resistor[{_node1_**,**_node2_**}, **_value_**, **_options_**]**

Specifies a simple resistor of _value_ Ω. _See also_ **ResistorThinFilm**.
Options: Temperature of resistor can be set using the rule
Temperature->tValue; **t**=300 Kelvin is the default.

ResistorThinFilm[{_node1_**,**_node2_**},** _ohmPerSquare_**, W, L, H, Kr,**
options**]**

Defines a resistor fabricated as a thin film of length _L_ and width _W_ from material
of resistivity _ohmPerSquare_ laid at a height _H_ (from ground plane) on a dielectric
material of value _Kr._
Options: Temperature of **ResistorThinFilm** can be set using
Temperature->tValue; **t** = 300 kelvin is the default.

Resistor types: _See_ the **Ribbon** and **Wire** in Section A.2.3.

SubNetwork[{_nodelist_**},** _specifier_**]**

Specifies a subnetwork connected at the nodes given in _nodelist_ that has an action
defined in the _specifier_ argument. _specifier_ must be a square list representation of a
_Y_-matrix or another matrix type. When the matrix elements are a function of frequency, they must use **f** as the frequency variable. **Draw[]** does not work with
SubNetwork[].

TransferDelay[{_node1_**,**_node2_**,**_node3_**,**_node4_**},** _t_**]**

Given a voltage (v_{In}) between _node2_(+) and _node1_, generates a voltage (v_{Out})
across _node4_ (+) to _node3_ where

$$v_{Out} = v_{In}e^{I2\pi ft}$$

TransferFunction[{_node1_**,**_node2_**,**_node3_**,**_node4_**},** _gain_**]**

Given a voltage (v_{In}) between _node2_(+) and _node1_, generates a voltage (v_{Out})
across _node4_(+) to _node3_ where $v_{Out} = v_{In}$ _gain_; _gain_ can be in terms of frequency
in which case **f** should be used as the frequency variable.

TransferPhase[{_node1_**,**_node2_**,**_node3_**,**_node4_**},**_phase_**]**

Given a voltage (v_{In}) between _node2_(+) and _node1_, generates a voltage (v_{Out})
between _node4_(+) and _node3_ where $v_{Out} = v_{In}e^{I\ phase}$

Transformer[{_node1_**,**_node2_**},** _n_**]**

Defines a simple two-winding transformer with an _n_:1 primary:secondary ratio.
The windings are between _node1_ and node 0 (primary) and _node2_ and node 0 (secondary).

Transistors: _See_ **BJT, FET, MESFET**, or **MOSFET**

TransmissionLine[{node1**,**node2**},** zo**,** theta**,** fo**]**

Specifies a transmission line that connects node1 to node2, of impedance zo, and phase change theta at frequency fo.

TransmissionLine4[{node1**,**node2**,**node3**,**node4**},** zo**,** theta**,** fo**]**

Specifies two transmission lines that connect node1 to node4 and node2 to node3, of impedance zo, and phase change theta at frequency fo.

TransmissionLineLossy[{node1**,**node2**},** zo**,** theta**,** fo**,** loss**,**
    options**]**

Specifies a transmission line that connects node1 to node2 of impedance zo, phase change theta at frequency fo, and loss in dB per meter.

Options: **Loss->Metal** includes a $\sqrt{f/f_o}$ dependence while **Loss->Dielectric** uses a f/f_o dependence. The temperature can be set using **Temperature->t**; **t** = 300 kelvin is the default.

TransmissionLineParalleCoupled[{node1**,** node2**,** node3**,**
    node4**},** zoe**,** thetae**,** zoo**,** thetao**,** fo**,** options**]**

Specifies a pair of coupled transmission lines for the nodal network.

Transmission Media Types: *See* the components of the family **TransmissionLine**, as well as **Coax**, **CoplanarWG**, **LineRLGC**, **Microstrip**, **MicrostripCoupledPair**, **ParallelWire**, **RectangularWG**, **Slotline**, **Stripline**, **StriplineCoupledPair**, **WireOverGround** in Section A.2.3.

VCCS[{node1**,**node2**,**node3**,**node4**},** gm**,** fc**]**

Defines a voltage-controlled current source; output current (from node4 to node3) is the product of gm (in mhos) and the voltage between node1(+) and node2. fc is optional and is the corner frequency for gm.

VCVS[{node1**,**node2**,**node3**,**node4**},** gain**,** fc**]**

Defines a voltage-controlled voltage source; output voltage at node4(+) is the product of gain and the voltage between node1(+) and node2. fc is optional and is the corner frequency for gain.

VoltageNoise[{node1**,**node2**},** value**,** options**]**

Specifies an rms voltage noise, per Hz.
Options: Temperature can be set using **Temperature->t**; **t** = 300 is the default.

VoltageSource[{_node1_**,**_node2_**},** _volts_**]**

Specifies a voltage source between _node1_ (+) and _node2_. If the value of _volts_ is a function of frequency, then **f** must be used as the frequency variable.

A.2.2 *Nodal* Functions

Nodal's functions cover three main areas of work: parameter calculation, noise parameter/correlation calculation, and production of graphics.

A.2.2.1 Parameter Calculation

There are five main parameter calculation functions that carry out the calculation of *ABCD*-, *Y*-, *Z*-, *S*-, and *T*-parameters. They all take either a nodal network or a matrix (giving parameters of some other type) as their first argument. The nodal network must be the result of the **NodalNetwork** function; the matrix must be an *S*-, *Y*-, *T*-, *Z*-, or *ABCD*-matrix, and the elements of the matrix must be in complex Cartesian form.

ABCDParameters[_firstArgument, options_**]**

YParameters[_netName, options_**]**

ZParameters[_netName, options_**]**

SParameters[_netName, options_**]**

TParameters[_netName, options_**]**

Options: To specify for which nodes parameter calculation is to be carried out use **Nodes -> {nodelist}**. For example, giving the option **Nodes -> {3,7}** instructs *Nodal* to carry out two-port analysis using node 3 as the input node and node 7 as the output node, both referenced to ground (node 0). When *Nodal* carries out calculations and displays results, it will map node number 1 onto node 3 and node number 2 onto node 7.

For nodal networks that contain components with frequency-dependent characteristics, you must either specify the frequency at which you want the calculation carried out or use the **f** variable in **Frequency->f** to signify that you want frequency left as a symbolic variable. For example, to analyze a circuit at a single frequency of 100 kHz, at three spot frequencies (10, 30, and 100 Hz), or leave frequency as a symbolic variable, you can use **Frequency->100 kHz**, **Frequency->{10,30,100}**, or **Frequency->f**, respectively.

Where you require the result of an analysis as an *s*-domain function, set **Frequency->Laplace**.

For *S*- and *T*-parameter calculation the nodal network is, by default, terminated at its input and output port by 50 Ω. If you wish to change the terminating impedance, you can use the option **Ports->z**, where **z** is the new impedance. For example, to terminate the ports with a 1 Ω resistance, use **Ports->1**. If the termination resistance is different for each port, then use a list to specify the terminating impedance (in the same order as the nodes are specified in the **Nodes** option): **Ports->{30,40}**.

NodalAnalyze[*netName, options*]

The nodal network to be analyzed is assigned to *netName*.
Option: The **Result->** option specifies what quantity you want as the result of the analysis; if no **Result->** option is given then the *Y*-parameters for the circuit are returned by default. **Result->***x*, where *x* is one of **Yij**, **Zij**, **Sij**, **Tij**, **Aij**, **Vk**, **Ik**, **NFmin**, **Rn**, **Yopt**, or **GammaOpt**; **i** and **j** are matrix element subscripts, and **k** is a node identifier.

For example, **Result->S21** calculates the forward transmission gain (S_{21} of the *S*-parameter family) of the circuit identified by the name *netName*. To calculate the voltage at node 5, you can use **Result->V5**. When currents are calculated (for example, using **Result->I5**), the node must be adjacent to a voltage source. (An alternative method for calculating the current through a circuit's branch is to insert a current-controlled voltage source in the branch and measure the voltage at the output of the source; the output of the source should not be connected to any other component.)

You can specify which nodes are output ports using the **Nodes->** option. **Nodes->{***nodeList***}** or **All**; *nodeList* must specify the input and output nodes, assuming node 0 is ground.

The frequency at which the analysis is to be carried out is specified by the **Frequency->** option: **Frequency->f** (to keep **f** as a symbolic variable), or **Frequency->{***frequencyList***}** for multiple frequencies, or **Frequency->** *f* for a spot frequency value.

A.2.2.2   Noise Parameter Calculation

CabcdParameters[*netName, options*]

CyParameters[*netname, options*]

CzParameters[*netName, options*]

Noise correlation matrices are returned for the network. These matrices are divided by 4 kT. The correlation matrices represent the equivalent noise sources at each network port. Conversions between the various noise parameters can be done by using

> **C?Parameters**[**C*Parameters**[*netName*],
> > ***Parameters**[*netName*]].

The **?** and ***** characters are wildcards for any parameter name.

NoiseParameters[*netName, options*]

This function returns rules describing the noise characteristics of the network. Minimum noise figure in dB and as a magnitude are returned. **Rn** is returned as well as **Yopt** and **GammaOpt**. **Yopt** and **GammaOpt** are the optimum source admittance and reflection coefficient, respectively. Options include the system characteristic impedance, **Zo**.

NoiseFigure[*sourceZ, netName*]

Returns the noise figure magnitude, *F*, for the network with the given source impedance.

A.2.3 *Nodal* Utilities

Nodal contains many utilities that perform minor calculations, in support of your circuit analysis. This section describes these utilities.

Ampere[Watt[w**], Zo->**zValue**]**

Returns the RMS current that results in a power dissipation of *w* watts when it passes through an impedance of *zValue*. If **Zo** is not specified, 50 Ω is taken as the default.

Angle[*complexNumber*]

Calculates the argument (in degrees) of *complexNumber*.

ArcDB[*dBvoltageRatio*]

Calculates the voltage ratio ($10^\wedge$(*dBvoltageRatio*/20)) that would result in a ratio, expressed in dB, of *dBvoltageRatio*.

ArcDBP[*dBpowerRatio*]

Calculates the power ratio ($10^\wedge$(*dBpowerRatio*/20)) that would result in a ratio, expressed in dB, of *dBpowerRatio*.

ArcDBRef[*dBpowerRatio*,**Reference->**refPower,**Zo->**zValue]

Calculates the power in watts that would result in a power ratio of *dBpowerRatio* being dissipated by an impedance *zValue* when the 0 dB reference level is *refPower*, in watts.

Capacitance[*l, cType*]

Returns the capacitance in farads of length *l* of conductor of type *cType*.

Coax[*innerR, outerR, dConstant* **]**

Defines a coaxial transmission line with inner conductor and outer shield diameters *innerR* and *outerR*, which are separated by a material of relative dielectric constant *dConstant*.

ComplexToPolar[*complexNumber* **]**

Converts a Cartesian $\{x + Iy\}$ *complexNumber* to a product of a magnitude and an exponential (which contains the phase information).

ComplexToPolarDBDegree[*complexNumber* **]**

Converts a cartesian $\{x + Iy\}$ *complexNumber* to a two-element list $\{$**DB[**complex-Number**]**, **Angle[**complexNumber**]**$\}$. The angle is given in degrees.

ComplexToPolarDegree[*complexNumber* **]**

Converts a cartesian $\{x + Iy\}$ *complexNumber* to a two-element list $\{$**Abs[**complex-Number**]**, **Angle[**complexNumber**]**$\}$. The angle is given in degrees.

CoplanarWG[*w,gapWidth,h,dConstant* **]**

Defines a coplanar waveguide transmission line of width *w*, separated from surrounding conductor by *gapWidth* at a height *h* above the substrate base, with the substrate having a relative dielectric constant *dConstant*
Option: **Thickness->***thick*; *thick* must be zero at present.

CoplanarWGinBox[*w,gapWidth,height,dConstant* **]**

Defines a coplanar waveguide transmission line of width *w*, separated from the surrounding conductor by *gapWidth* at a height *h* above the substrate base, with the substrate having a relative dielectric constant *dConstant*. *dConstant* is optional and, if omitted, is unity by default.
Option: **Thickness->***thick*; *thick* must be zero at present.
Option: **CoverHeight->***hc*; *hc* is the distance between the waveguide and the grounded cover.

DB[*ratio* **]**

Returns the dB equivalent for a voltage or current *ratio*.

DBP[*powerRatio* **]**

Returns the dB equivalent for a power *ratio*.

DBRef[Volt[*level***]]**

Returns the dB equivalent for a signal with voltage *level*, referenced to a power level of 1 mW into an impedance of 50 Ω. The signal can also be specified in amperes and watts by using **Ampere[***level***]** and **Watt[** *level***]**.
Option: The reference power level can be specified to a value other than 1 mW by **Reference->***refPower.*
Option: The impedance into which the power is fed can be specified to be a value other than 50 Ω by **Zo->***zValue.*

Delay[*l, lineDescription***]**

Calculates the delay in seconds for a line *l* meters long with either a *lineDescription* set to the relative dielectric constant or to one of the transmission media types (for example, **MicroStrip**). If no *lineDescription* is specified, then a unit dielectric constant is assumed.

DielectricConstant[*TransmissionMediaType***]**

Returns the effective dielectric constant of the given *TransmissionMediaType.*
Option: **Frequency->***fValue* for dispersive media.

Draw[*componentName[ ]***]**

Displays a picture of the device *componentName[ ].*

FourierEE[*timeSeries***]**

Computes the single-sided spectrum of *timeSeries.*

Gap[*w, h, dConstant***]**

Defines a capacitive gap of width *w* and height *h*. The gap filler is of relative dielectric constant *dConstant*; if *dConstant* is omitted, it is assumed to be unity.

Inductance[*l, ConductorType***]**

Returns the inductance of the given *ConductorType*, *l* meters long.

Interdigital[*w, gap, n, dConstant***]**

Specifies an interdigital capacitor structure that has *n* fingers each of width *w* and finger-to-finger separation *g*, on a substrate of effective dielectric constant *dConstant*. If *dConstant* is not given then it is assumed to be unity.

InverseFourierEE[*singleSidedSpectrum*]

Returns the time-domain representation of the *singleSidedSpectrum*.

LineRLGC[*r*,*l*,*g*,*cap*]

Defines a general per-unit line, with parameters of resistance *r*, inductance *l*, characteristic impedance *g*, and capacitance *cap*, given per meter.
Option: **Frequency->***fValue*

MatrixColumnForm[*matrixList*]

Formats a list of matrix data types, *matrixList*, as a column of matrices.

Microstrip[*w*,*h*, *dConstant*]

Defines a microstrip with width *w*, on a substrate of height *h* and relative dielectric constant *dConstant*. *dConstant* is optional and is assumed to be unity if omitted.
Option: The strip thickness can be specified using the option **Thickness->***thick*. Default value is *thick*=0.
Option: Frequency effects can be specified using the option **Frequency->***fValue*

MicrostripCoupledPair[*w*, *g*, *h*, *dConstant*]

Defines a microstrip coupled pair of lines, each of width *w*, separated by a gap of size *g*, on a substrate of height *h* and dielectric constant *dConstant*. If *dConstant* is omitted, it is assumed to be unity.

Pad[*attn*]

Returns the resistor values for a pad with attenuation *attn* (measured in dB).
Option: The impedance of the pad can be specified using the option **Zo->***zValue*; **Zo**=50 Ω by default.
Option: The geometry of the pad can be specified using the option **Type->Pi** or **T**; **Pi** is the default.

Parallel[*res1*,*res2*]

Combines the two arguments as though they were resistors to give the result *res1 res2*/(*res1+res2*).

ParallelPlate[*w*,*h*, *dConstant*]

Defines a parallel plate capacitor structure of width *w*, height apart *h*, and with a separating medium relative dielectric constant *dConstant*. If *dConstant* is omitted, it is assumed to be unity.

ParallelWire[*wDiameter*, *s*, *dConstant*]

Defines a transmission medium type of two parallel wires, each of diameter *wDiameter*, separated by a distance *s* in a medium of relative dielectric constant *dConstant*. If *dConstant* is omitted it is assumed to be unity.

PhaseShift[*f*, *l*, *lineDescription*]

Returns the phase shift for a signal of frequency *f*, traveling along a line of length *l*, and described by *lineDescription* where *lineDescription* is either a relative dielectric constant or a transmission medium type. If *lineDescription* is omitted then it is assumed that the dielectric constant is unity.

PolarDBDegreeToComplex[{*db*, *angle*}]

Converts the polar coordinate vector given by the number pair {*db*, *angle*} to a complex number. *db* is in decibels, and *angle* is given in degrees.

PolarDegreeToComplex[{*r*, *angle*}]

Converts the polar coordinate vector given by the number pair {*r*, *angle*} to a complex number where *angle* is given in degrees.

PolarToComplex[*r* **E^(I** *theta***)**]

Converts the polar coordinate vector given by $r E^{I\,theta}$ to a Cartesian complex number where *theta* is given in radians.

RectangularWG[*l*, *w*, *dConstant*]

Defines a rectangular waveguide transmission media-type object of length *l* and width *w* and relative dielectric constant *dConstant*. If *dConstant* is omitted then it is assumed to be unity.
Option: The mode can be set using **Mode->**modeName, where *modeName* is either **TEmn** or **TMmn**.
Option: **Frequency->**fValue sets the evaluation frequency to *fValue*.

Resistance[*l*, *cType*]

Returns the resistance in ohms of a conductor of type *cType* and length *l*.
Option: Setting **Frequency->**fValue includes frequency effects.

Ribbon[*w*, *thick*]

Defines a ribbon structure of width *w* and thickness *thick*.

Options: The height above the ground plane is assumed to be infinite but can be specified by setting **Height->***h*. The resistivity of the ribbon can be specified by **Rho->***resistivity*.

SingleLayerSolenoid[*d*, *nTurns***]**

Defines a single-layer solenoid inductor of diameter *d* with *nTurns* turns of wire.

SkinDepth[*r*, *f***]**

Calculates the skin depth of material with resistivity *r* at a frequency *f*.

Slotline[*w*, *h*, *dConstant***]**

Defines a slotline transmission media type containing a gap of width *w*, on a substrate of height *h* and dielectric constant *dConstant*. If *dConstant* is omitted it is assumed to be unity.
Options: Conductor thickness is set by **Thickness->***thick* and the evaluation frequency by **Frequency->***fValue*.

Solenoid[*innerD*, *outerD*, *nTurns***]**

Defines a multilayer solenoid with inner and outer diameters *innerD* and *outerD*, with *nTurns* as the number of turns. (*See also* **SingleLayerSolenoid**, **Spiral** and **Toroid**.)

Spiral[*innerR*, *w*, *g*, *h*, *nTurns***]**

Defines a spiral inductor defined by an inner radius *innerR*, a conductor width *w*, and interspiral gap *g*, and with *nTurns*. The height above the ground plane is *h*.

Stripline[*w*, *g*, *dConstant***]**

Defines a stripline transmission media type of width *w*, ground spacing *g*, in a medium of dielectric constant *dConstant*. If *dConstant* is omitted it is assumed to be unity.
Options: **Thickness->***tValue* sets the conductor thickness.

StriplineCoupledPair[*w*, *g*, *gs*, *dConstant***]**

Defines a coupled pair of striplines each of width *w*, ground spacing *g*, in a medium of dielectric constant *dConstant*. If *dConstant* is omitted it is assumed to be unity.
Options: **Thickness->***tValue* sets the conductor thickness.

Toroid[*innerD, outerD, wireD, nTurns***]**

Defines a toroidal inductor of inner and outer diameters *innerD* and *outerD*, wound with *nTurns* turns of wire of wire diameter *wireD*.
Options: **Permeability->***pValue* sets the toroid's relative permeability.

VelocityOfLight[*dConstant***]**

Returns the velocity of light in a medium with dielectric constant *dConstant*.

ViaHole[*d***]**

Defines an inductor created by a via hole of diameter *d*.
Options: Set **Method->Pucel** or **Tripathi** to specify the calculation model (**Pucel** is the default). The **Tripathi** method requires that the hole metallization thickness be specified. This thickness is set using **MetalThickness->***thick*. **MetalThickness->d/10** is the default.

Volt[Watt[*p***]]**

Returns the RMS voltage of a signal which dissipates a power *p* in an impedance of 50 Ω .
Options: Use **Zo->***zValue* to set difference impedance values.

Watt[Volt[*v***]]**

Returns the power dissipated by a voltage *v* in an impedance of 50 Ω. You can use current instead of voltage by specifying **Ampere[***i***]**.
Options: Use **Zo->***zValue* to set difference impedance values.

Wavelength[*f, lineDescription***]**

Returns the wavelength of a signal of frequency *f* traveling in a medium or along a transmission line. For a medium, *lineDescription* is the relative dielectric constant. Use a member of the transmission media type to specify a transmission line.

Wire[*d***]**

Defines a wire of diameter *d*.
Options: The height above the ground plane can be specified by **Height->***h*. The resistivity, *r*, of the wire's material is set using **Rho->***r*.

WireOverGround[*d, h, dConstant***]**

Defines a wire over a ground plane. Wire diameter is *d*, the height above the ground plane is *h* in a medium of relative dielectric constant *dConstant*. If *dConstant* is omitted it is assumed to be unity.

Zo[*tmt***]**

Returns the characteristic impedance of the transmission media type *tmt*.
Options: Frequency of evaluation can be set using the option
Frequency->*fValue*.

A.3 Graphics

Nodal provides graphical functions that are especially useful in engineering, over and above the rich selection of graphical functions with *Mathematica. Mathematica* is very flexible about the presentation of graphics. But, if you are new to *Mathematica*, this flexibility can be rather confusing and you will be rewarded by a browse through the graphics sections of both *Mathematica — A System for Doing Mathematics by Computer* and Wolfram Research's Technical Report *Guide to Standard Mathematica Packages* (both of which come with *Mathematica*).

In this section we give a brief guide to plotting different types of data, including commonly used options for each command. For many *Mathematica* functions, we merely list the option or function names that you can use to achieve the desired result (since there are so many combinations of graphics functions in *Mathematica*, a full exposition of the subject is outside the scope of this book).

A.3.1 Lists

You can use **ListPlot** to create graphs of a list of isolated *y*-values (with implicit assumption of equal spacing in the *x*-coordinate), or a list of {*x*, *y*} pairs, or **ListPlot3D** for {*x*, *y*, *z*} triples:

> Labeling axes: **AxesLabel->{"***xlabel***","***ylabel***"}**
> Labeling plot: **PlotLabel->"***text***"**
> Choosing axes intersection: **AxesOrigin->{***xVal, yVal***}**
> Selecting points or line: **PlotJoined->True|False**
> Altering point/line thickness: *See* **PointSize**.
> Plots with error bars: *Use* **ErrorListPlot**.

In:
```
yList={1,3,5,7,5,3};

xyList={{10,1},{12,3},
        {12.5,5},{15,7},
        {16,5},{16.5,3}};

ListPlot[yList,
        AxesOrigin->{0.5,0},
        PlotRange->{{0,8},{-2,10}},
        PlotJoined->False,
        AxesLabel->{"x","y"},
        PlotLabel->"my data"];
```

Out:

The next example demonstrates a scatter graph and changes the font used for tick and title labeling. The data for the scatter graph, **xyList**, are in the form of a list of pairs. (If your data were generated as two separate lists, you can convert them to a list of pairs using **Transpose**: See the following **LogListPlot** example.)

In:
```
ListPlot[xyList,
        PlotJoined->True,
        DefaultFont->{"Times-Bold",12},
        PlotLabel->FontForm["Title",
                {"Helvetica-Oblique",10}]];
```

Out:

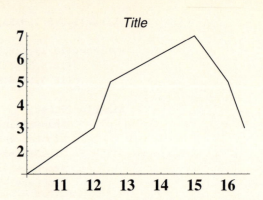

When some feature appears in your data, you may want to label it with text. You can place text anywhere in a diagram and specify its reading direction as well as font and face by using the graphics primitives for text. In the following example, you label the point {11,2} with the text "SA" in normal face and Times 14 point as the font. The third argument of **Text** specifies that the lower left-hand corner of the box, which holds the text, is to be placed at {11,2}; the last argument specifies that the text is to read from top to bottom.

The last two (positioning) arguments of **Text** allow you to place text so that it does not (or does!) overlap the feature on the graph to which it is referring.

In:
```
Show[%,
     Graphics[Text[FontForm["SA",
                            {"Times",14}],
              {11,2},
              {-1,-1},
              {0,-1}]]];
```

Out:

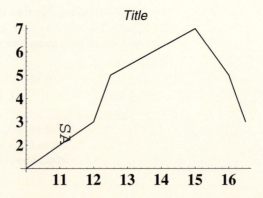

A wide range of functions are available for specifying color and grayscale attributes for the background, text, and other elements in your plot. You can also include text or graphics objects in plots with the **Epilog** or **Prolog** options, as shown in the following section.

A.3.2 Log Plots

The **Graphics`Graphics`** package contains the functions **LogListPlot** and **LogLogListPlot**, which enable you to plot list data with y-axis, or x- and y-axes, logarithmically scaled.

In:
```
Needs["Graphics`Graphics`"];

xdata={0.01,0.1,1.0,10.0,100.0};
ydata={1,3,10,30,100};
LogListPlot[Transpose[{xdata,ydata}]];
```

Out:

You can issue directives that refer to graphics primitives that your graph will use. One way to change the point size that will be used by *Mathematica* when it plots points is to issue the **PointSize** directive as part of the setting up work (the **Prolog**) that *Mathematica* performs during execution of the graphics function. Point size, line thickness, and dashing can all be specified in either relative or absolute units. An absolute size directive prescribes the actual size of the object on the display (by default, in points), whereas the relative size directive specifies object sizes as a fraction of the total width of the plot. Both are very useful when you want to plot diagrams that will be subsequently reduced in size — specifying object sizes prevents thin lines or small points disappearing during the reduction process.

In:
```
LogLogListPlot[Transpose[{xdata,ydata}],
              GridLines->Automatic,
              Prolog->PointSize[0.03]];
```

Out:

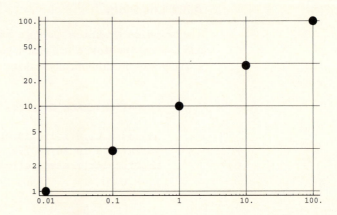

A.3.3 Adding a Legend

The **Graphics`Legend`** package adds options to the plotting commands that
give you control over legend plotting, placement, content, and so on. You can
style each line in terms of its color, thickness, and size of dashes in dashed lines.
Both the **PlotStyle** and **PlotLegend** options take a list as their argument.
The order of elements in this list corresponds with the order of functions to be
plotted in the **Plot** function's first argument:

In:
```
Needs["Graphics`Legend`"];

{top,bottom} = {1,-1};
{left,right} = {-1,1};

Plot[{x,x^2},
     {x,-2,2},
     PlotStyle->{Dashing[{.01}],
                 Dashing[{.02}]},
     PlotLegend->{"x","xSquared"},
     LegendPosition->{right,bottom}];
```

Out:

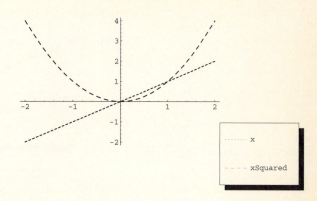

A.3.4 Multiline Plots

You can plot multiple *y*-values using either *Mathematica*'s **MultipleListPlot** function (which is contained in the **Graphics`MultipleListPlot`** package) or *Nodal*'s **NodalPlot** function.

MultipleListPlot expects a list of {*x,y*} pairs for each line to be plotted. **NodalPlot** works with either a list of coordinate sets, {x_1, y_{1a}, y_{1b}, y_{1c}}, like a spreadsheet, or a list of lists of {*x,y*} pairs for each line similar to **MultipleListPlot**. The following example shows how to use **NodalPlot** and **MultipleListPlot**:

In:
```
Needs["Graphics`MultipleListPlot`"];

xdata={1,3,4,5,8,9};
y1data={1,1,2,1,1,1};
y2data={5,4,3,6,5,4};
y3data=Table[Sin[x],{x,1,6}] //N;
line1=Transpose[{xdata,y1data}]
```
Out:
```
{{1, 1}, {3, 1}, {4, 2}, {5, 1}, {8, 1}, {9, 1}}
```

In:
```
line2=Transpose[{xdata,y2data}];
line3=Transpose[{xdata,y3data}];

MultipleListPlot[line1, line2, line3,
                PlotJoined->True];
```

Out:

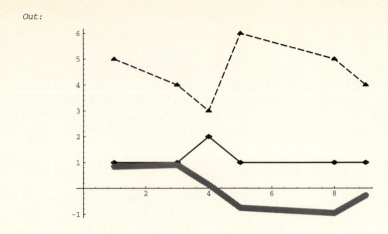

In:

```
Needs["Nodal`","Nodal2.m"];

NodalPlot[{line1,line2,line3},
        Label->{"line a","line b","Sin"}];
```

Out:

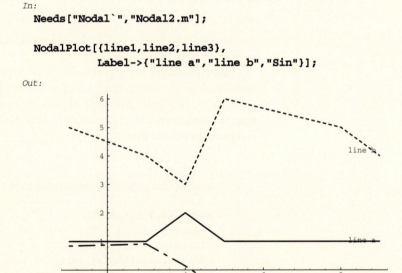

A.3.5 Magnitude and Phase Plots

The output from many forms of circuit analysis describes both the magnitude and phase of one or more signals. Magnitude and phase can be well shown in a polar plot (which we describe in Section A.3.6), but polar plots are often difficult to interpret, perhaps just because they are not as common as their *x-y* Cartesian counterparts.

The Bode plot is a convenient and widely used plot for the display of magnitude and phase information. *Nodal* provides you with a **BodePlot** function that is designed to display simultaneous magnitude and phase information.

BodePlot's first argument is a list of complex numbers or a list of complex number pairs, or a set of such lists, depending on the number of lines you wish to draw. Here is a simple example that plots a list of number pairs. Each pair consists of an *x*-value (which will normally be the frequency in Hz) and a (complex) *y*-value representing the magnitude and phase that you want to plot.

In:
```
b={{1,1},{10, -I},{100, -1-I}};
BodePlot[b,
          AxesLabel->{"frequency","gain (dB)"},
          Label->{"magnitude","phase"}];
```

Out:

Setting **Resolution->Course** plots the grid at decade and half-decade intervals only.

BodePlot is set up to make plotting of results as easy as possible. When you use it to plot output from **NodalAnalyze**, *Nodal* sets the axes labels and scales for you.

In:
```
circ=NodalNetwork[VoltageSource[{1,0},vIn],
                   Resistor[{1,2},10 kOhm],
                   Capacitor[{2,0},1 uF]];
result=NodalAnalyze[circ,
            Result->V2/V1,
            Step->Frequency[10^Range[0,5]]];

BodePlot[result];
```

Out:

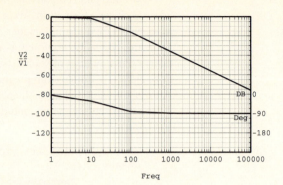

A.3.6 Polar Plots

Sometimes you will want to plot data that is more meaningful in a polar graph —
for example, the beam pattern of an antenna. *Mathematica* has a family of functions that plot polar data (radius as a function of angle) on a Cartesian grid, and
Nodal adds the command **NodalPolar**, which uses a polar grid.

Mathematica's polar-plotting functions are in the standard package
Graphics`Graphics`. By analogy with the Cartesian Plot family of
functions, **Plot[***f***, {***t***,***tmin***,***tmax***}]** plots the radius resulting from the function
f as the angle *t* varies from *tmin* to *tmax*. **PolarListPlot** takes a table of radii and
plots them at equispaced angles:

In:
```
rList=Table[r, {r,0,2 Pi,0.1}] //N;
PolarListPlot[rList];
```

Out:

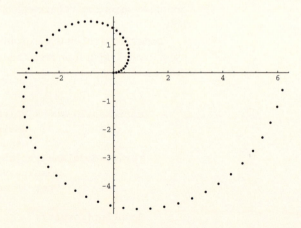

NodalPolar is principally intended to support the plotting of (relative) quantities as a function of phase (for example, S-parameter and Smith charts), but, of course, you can use it for plotting any polar data. **NodalPolar** expects a list of complex numbers. If your data are in polar form, $\{r, \theta\}$, you can use *Nodal*'s **PolarToComplex** utility to convert them into the Cartesian complex form, $\{x, I\, y\}$. The outer circle of **NodalPolar**'s plot represents the unit circle, $r = 1$ for all θ. By default, **NodalPolar** labels lines using S-parameter names. To override the default names, you can use the **Label->** option, as shown in the next example.

It is often difficult to distinguish the beginning from the end of a line drawn on a polar graph. To overcome this problem, **NodalPolar** has an option, **Ends->**, which can be set to **True** or **False**. If you set **Ends->True** then a dot and arrow will be placed at the beginning and end of the line, respectively:

In:
```
rList = Table[
            PolarDegreeToComplex[{r/(2 Pi),
                                  180r/Pi}
            ],
            {r,0,2 Pi,0.1}] //N;
NodalPolar[rList,
            Label->{"rList"},
            Ends->True];
```

Out:

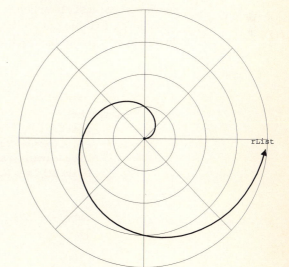

If you want to plot discrete points rather than lines, you can switch the option **PlotJoined->False** and then set the option **PointLabel** to **True** to switch on markers at each point. You must also specify the marker format using the option **PointStyle->{**_description_**}**, where _description_ is an anonymous function that will accept a list of number pairs, each pair representing the coordinates of the point to which the marker specified in _description_ will be attached. Since **PointStyle** is expected to cope with multiple lines, you must supply a list of styles (one style for each line to be drawn), enclosed in **{}**, even if the list is only one item long. For example, setting **PointStyle -> {Text["x",#]&}** will plot an **x** at each point. For two lines, plotted with symbols **x** and **o**, **{Text["x",#]&, Text["o",#]&}** would be used.

You can also use graphics primitives to mark each point. In the following example, you plot **rList** again, but this time you mark each position with a circle whose radius is a function of the radial distance from the center of the circle to the point being marked. When **NodalPolar** is executed, the anonymous function you must define for use by the **PointStyle** option is fed the coordinates of each point, in turn, in the form of a pair ($\{x,y\}$). The **Circle** primitive takes two arguments. The first is the coordinates of the circle's center, and you can use the **#** (**Slot**) function to copy the coordinate pair supplied by **NodalPolar** into **Circle**'s first argument. **Circle**'s second argument (in this form of the function) is a number that specifies the radius of the circle to be drawn. For this example, you want the radius to equal one-fifth of the radial distance of the point from the plot's center. You can calculate the radius using the Pythagorean theorem since you know the _x_- and _y_-coordinates of the point: **#[[1]]** and **#[[2]]**.

#[[1]] takes the first of the pair of numbers, the _x_-coordinate, which is supplied to the **PointStyle** option by **NodalPolar** and **#[[2]]** takes the _y_-coordinate. Both are squared, added together, and then the sum's square root is divided by **5** to give us the required radius for plotting by **Circle**.

```
In:
NodalPolar[
        rList,
        PointLabel->True,
        PlotJoined->False,
        PointStyle->{Circle[#,
                        Sqrt[#[[1]]^2 +
                                #[[2]]^2]/5]&},
        Label->{"A"}
        ];
```

Out:

A.3.7 Smith Charts

The Smith chart plotter has two forms: `SmithChart` and `SmithPolar`. Each of these plots works just like `NodalPolar`. The background for the `SmithPolar` plot is a `SmithChart` on the lower half and a polar chart on the upper half. This hybrid chart is very useful for plotting transistor S-parameters.

A.4 Importing Data

To think of *Mathematica* as a purely theoretical tool, never having any contact with the real world, would be very wrong. *Mathematica* can help you analyze and plot results from experimental work, too. The integrated analysis, modeling, and graphing environment is, we think, the ideal helper to have on your lab bench.

At the heart of getting *Mathematica* to interact with your data is the ability to read from and write to files. In this section, we describe *Mathematica*'s input/output functions. Because data and files can be in so many different formats on different computers, *Mathematica*'s functionality is not as sophisticated as software might be that just had to run on one machine, but we think being able to move your own *Mathematica*-based work to different machines is good compensation.

Learning how to read and write files involves learning to move around the directory structure of your computer, learning to open and close files, and understanding how *Mathematica* can treat different forms of file contents.

To find out in which directory *Mathematica* is currently looking, you can use the **Directory** command. The exact form of output is machine dependent — the examples here were executed on an Apple Macintosh — but the general principles apply to all machines on which *Mathematica* works. If you try to read from a file that is not in the current directory, *Mathematica* will return an indication of failure:

In:
```
Directory[]
```

Out:
```
Q127Alpha:Mathematica 2.2
```

In:
```
ReadList["datafile"]
```

Out:
```
$Failed
```

To set the current directory to the one containing your data, you can use the **SetDirectory** command. You can then use **FileNames** to confirm the directory contents:

In:
```
SetDirectory["Q127Beta:Sam's folder"]
```

Out:
```
Q127Beta:Sam's folder
```

In:
```
FileNames[]
```

Out:
```
{datafile, puredata}
```

If your file contains only numbers, separated either by spaces or placed on new lines, *Mathematica* can easily read in all the numbers as a list. The function **ReadList** takes two arguments: the name of the file from which the data are to be read and the type of data to be received. For example, a file named "**puredata**" contains numbers (both integer and real) that can be read into a list named **myData**:

In:
```
myData=ReadList["puredata",Number]
```

Out:
```
{1, 2, 23, 5, 8, 1.1, 1.2, 1.3, 1.5, 2.1, 2.2}
```

If you want the data read in as individual characters, then you need to specify **Character** as **ReadList**'s second argument.

```
In:
ReadList["puredata",Character]
```

```
Out:
{1,  ,  , 2,   , 2, 3,  , 5,  , 8,  , 1,  ., 1,  , 1,  .,
2,  , 1,  ., 3,  , 1,  ., 5,  , 2,  ., 1,  , 2,  ., 2}
```

The resulting list contains all the characters in the file, each as an individual element in the list. By specifying **Byte** as the second argument, the list would contain all the bytes in the file, expressed in the list as base-10 integers.

Note that each element (specified by **ReadList**'s second argument) has been read in individually and becomes an element in the resulting list. You may sometimes want to keep numbers (or a mix of numbers, strings, and so on) in some kind of structure. For example, if the numbers in your data file are in *x-y* pairs, and it is useful to keep that structure, then you can directly use a list of pairs in **ListPlot**. Maintaining structure is very simple. For example, to read in pairs of numbers, use **ReadList["filename", {Number,Number}]**.

If your data contain a mixture of text and numbers, you may wish to read in the data as a string and then search for text that indicates the presence of a subsequent number. For example, an avionics package under test is linked to an IEEE-488 logger that returns parameters as strings. Here is a printout of part of the logger's file that you have read in as a list of words:

```
In:
inString=ReadList["flightlog",Word]
```

```
Out:
{Altitude, 310, Offset, 10, mV, Drift, 100, mA,
Time, +09:20;Altitude, 310, Offset, 10, mV,
  Drift, 100, mA, Time, +09:21;Altitude, 315,
  Offset, 11, mV, Drift, 100, mA, Time,
  +09:22;Altitude, 315, Offset, 10, mV, Drift,
  100, mA, Time, +09:23;Altitude, 315, Offset,
  11, mV, Drift, 90, mA, Time, +09:24;Altitude,
  320, Offset, 10, mV, Drift, 100, mA, Time,
  +09:25;Altitude, 320, Offset, 12, mV, Drift,
  90, mA, Time, +09:26;Altitude, 325, Offset, 10,
  mV, Drift, 100, mA, Time, +09:27;}
```

If you want to plot all the offset values then one way of doing this is to note that a value always follows the word "Offset".

```
In:
offsetList={};
For[i=1, i<Length[inString], i++,
    If[inString[[i]]==="Offset",
        AppendTo[offsetList,
                inString[[i+1]]]]]
```

You can then easily examine the contents of **offsetList**:

In:
```
offsetList
```

Out:
```
{10, 10, 11, 10, 11, 10, 12, 10}
```

Mathematica will not be able to plot or operate numerically on this list, because its elements are actually strings — as indicated by the quote marks in the full form of one of **offsetList**'s members:

In:
```
FullForm[offsetList[[1]]]
```

Out:
```
"10"
```

To use the elements of **offsetList** numerically, you must convert them from strings to numbers. *Mathematica* contains a wide variety of functions for manipulating strings, including conversion from string to number and vice versa. You can map the function **ToExpression** onto the list of strings and so convert it into a list of *Mathematica* expressions, which can then be evaluated in the usual way:

In:
```
ToExpression /@ offsetList
```

Out:
```
{10, 10, 11, 10, 11, 10, 12, 10}
```

In:
```
ListPlot[%,
         PlotJoined->True,
         PlotRange->{8,14}];
```

Out:

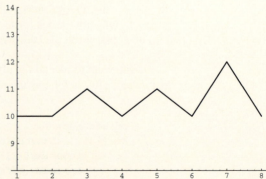

When converting from strings to numbers, remember that only **ReadList** will correctly interpret BASIC, C, or FORTRAN format numbers, which have e or f (or any other alphabetic character) to indicate the beginning of the exponent. Representing 10000 as **1e4** is incorrect *Mathematica* syntax: **ToExpression** will treat **1e4** as the product of **1** and the variable **e4**. But you can use the **StringToStream** function to convert the string **a** into an input stream, which can then be digested by **ReadList**:

```
In:
  a="1e4";
  ToExpression[a]

Out:
  e4
```

```
In:
  ReadList[StringToStream[a],Number]

Out:
  {10000}
```

You need not worry about whether your input data contain the correct case. The functions **ToUpperCase** and **ToLowerCase** will make all characters in a string upper- or lower-case, respectively. To help you with parsing a string, *Mathematica* also provides the tests **DigitQ**, **LetterQ**, **UpperCaseQ**, **LowerCaseQ** to check for all characters in a string being numerical, alphabetic, upper-case, and lower-case, respectively.

You may find that sometimes it is easier to treat all the text in your data file as one string, rather than a sequence of, say, words or characters. Doing so opens up new ways of elegantly processing file contents. The next example rereads the data from the logger as one string and then extracts all the altitude values for plotting. After you read in the file contents, you have to find the altitude values (which are conveniently sited after the ASCII string "Altitude"), extract each of them, so forming a list of strings with each element being one value. First, the file is read in using **ReadList**, and then the start and end position of each occurrence of the string "Altitude" is found. (You are not really interested in where that string begins, but **StringPosition** gives us both start and end positions. The **** symbol is *Mathematica*'s line continuation indicator.)

```
In:
  input=ReadList["flightlog",String]

Out:
  {Altitude 310 Offset 10 mV Drift 100 mA Time\
     +09:20;Altitude 310 Offset 10 mV Drift 100 mA\
     Time +09:21;Altitude 315 Offset 11 mV Drift\
     100 mA Time +09:22;Altitude 315 Offset 10 mV\
```

```
         Drift 100 mA Time +09:23;Altitude 315 Offset\
         11 mV Drift 90 mA Time +09:24;Altitude 320\
         Offset 10 mV Drift 100 mA Time +09:25;Altitude\
         320 Offset 12 mV Drift 90 mA Time\
         +09:26;Altitude 325 Offset 10 mV Drift 100 mA\
         Time +09:27;}
```

In:
  posList=StringPosition[input[[1]],"Altitude"]

Out:
  {{1, 8}, {52, 59}, {103, 110}, {154, 161},
    {205, 212}, {255, 262}, {306, 313}, {356, 363}}

You used **input[[1]]** not just **input**, to refer to the string created by **ReadList**, because **ReadList** returns a list of strings — albeit one element long in this case — enclosed in **{}**. The characters containing the altitude value are sited in the four positions which follow the string "Altitude". As a simple way of getting at the data you want, you can use the function **StringTake** to extract characters from a large string. However, before you can use **StringTake**, you need a list that contains the start and end positions of the strings you wish to extract from **input**. To create this list, you can define an anonymous function that adds **1** and **4** to the end position of all the "Altitude" strings reported in **posList**, and then map (**/@**) it onto each element of **posList**. This process produces a list of the start and end positions of each altitude value, called **getList**:

In:
  getList=({1+#[[2]],4+#[[2]]})& /@ posList

Out:
  {{9, 12}, {60, 63}, {111, 114}, {162, 165},
    {213, 216}, {263, 266}, {314, 317}, {364, 367}}

By mapping the function **StringTake** over **getList**, you extract all the strings containing the altitude values:

In:
  (StringTake[input[[1]],#])& /@ getList

Out:
  {310, 310, 315, 315, 315, 320, 320, 325}

The result is just a list of strings, of course. If you want the altitude values as numbers, you can use **ToExpression**:

In:
  ToExpression[(StringTake[input[[1]],
                  #])& /@ getList]

Out:
  {310, 310, 315, 315, 315, 320, 320, 325}

To plot a graph of offset values as a function of altitude, you can convert the list of offsets to numbers, too:

In:
  ToExpression /@ offsetList

Out:
  {10, 10, 11, 10, 11, 10, 12, 10}

Since **ListPlot** requires a list of {*x,y*} pairs — and not two lists of *x*- and *y*-values — you can rearrange your data using **Transpose** before plotting. (The altitude axis is the *y*-axis because it seems more realistic!)

In:
  ListPlot[Transpose[{%,%%}],
  **   PlotJoined->False,**
  **   PlotRange->{{9,13},{300,350}}];**

Out:

A.5 Exporting Data

You can export data from *Mathematica* to external files. The basic sequence of operations is the same one you used for reading in data. First, you must set a directory into which the data will be written. Once the directory has been specified with the **SetDirectory** function, you can use either the **OpenWrite** or **OpenAppend** functions. These both take a filename, as a string, as their argument and both open a file with that name, but **OpenWrite** gives you a new, empty file (that is, it will overwrite any extant file of the same name); **OpenAppend** will append any new output onto any extant file. **OpenAppend** is safer; if you want to check which files are in the directory you are about to use then the **FileNames** function will list them. In the following example, you create a new file, write a table of values to it, and finally write the string assigned to the variable **text**. **outStream** is the i/o stream identifier (which will be familiar to C programmers). Stream identifiers are required because you can have more than one stream open at a given time: By providing a list of stream identifiers as the stream argument of the **Write** family of functions, you can write to multiple streams with one command:

In:
```
SetDirectory["Q127Beta:Sam's folder"]
```
Out:
```
Q127Beta:Sam's folder
```

In:
```
outStream=OpenWrite["output"]
```
Out
```
OutputStream[output, 5]
```

In:
```
Write[outStream,Table[{i,Tan[i]//N},{i,1,10}]]
text="end of my data";
WriteString[outStream,text]
```

If you want to insert tabs or newline characters, you can use their string form in *Mathematica*: **\t** or **\n**. Any 8-bit value can be inserted into a string using the form **\.hh**, where **hh** is the hexadecimal value of the character to be inserted; 16-bit characters, used for Japanese characters, for example, are entered using **\.hhhh** by extension of the technique.

Once you have finished writing to a stream, you should use the **Close** function to formally terminate use of that stream. By using a text editor on your computer, you can verify the file contents:

In:
```
Close[outStream]
```
Out
```
output
```

On the Apple Macintosh, using TeachText (a simple text editor supplied with all Macintosh machines) enables you to see the contents of the file just created:

```
{{1, 1.557407724654902225}, {2, -2.185039863261518996},
 {3, -0.1425465430742778052}, {4, 1.15782128234957759},
 {5, -3.380515006246585631}, {6,-0.2910061913847491586},
 {7, 0.8714479827243187416}, {8, -6.79971145522037868},
 {9, -0.4523156594418098416}, {10,0.648360827459086676}}
end of my data
```

Mathematica can also format your data for use with spreadsheets and other programs. The basic idea is to use **OutputForm** to force **Write** to output formatted data. The following code uses the preceding data but formats the output file. We format the data in two columns with **TableForm** and then we use the **TableSpacing** option to set no lines between the data and three spaces between the columns. The code for writing this new output file, "**output2**", is given below.

In:
```
outStream=OpenWrite["output2"];
Write[outStream, OutputForm[
    TableForm[Table[{i,Tan[i]//N},{i,1,10}],
        TableSpacing->{0,3}]]];
text="end of my formatted data";
WriteString[outStream,text];
Close[outStream]
```

Out
```
output2
```

The file named "**output2**" contains the following information:

```
1    1.55741
2    -2.18504
3    -0.142547
4    1.15782
5    -3.38052
6    -0.291006
7    0.871448
8    -6.79971
9    -0.452316
10   0.648361
end of my formatted data
```

A.6 Example Code Usage

This section contains the definitions for the main functions developed in this book. The accompanying disk also contains usage statements for the example code. Function names beginning with a lower-case letter are usually used for a specific and often temporary application. Function names beginning with a capital letter are more general and are intended to be used in many applications. The **[..]** notation refers to a variable or several variables separated by commas.

ABCDData [*matrix*]

Is a wrapper for ABCD matrix data. **ABCDData** works just like **ABCDMatrix** in *Nodal* but a different name was used to avoid conflicts.

ABCDSeries [*z*]

Returns an ABCD two-port matrix representing the series element, *z*.

ABCDShunt [*y*]

Returns an ABCD two-port matrix representing the shunt element, *y*.

Amplifier [..]

A wrapper that holds zero or more rules for amplifier Gain, NF, and/or IP3. The rules are separated by commas. Defaults are given in the options for **Amplifier**.

arma [*list*]

Returns the data filtered by a *z*-plane pole and zero given in Section 8.5.3. **arma** is fast and easily modified for your own application.

Cascade [..]

Is a wrapper which holds **Amplifier** data types separated by commas. **Cascade** is used by **Tabulate** and **Draw**.

CascadeForm [..]

Is the wrapper for Cascade system data output. **CascadeForm** is used by **PlotCascade**.

CFELadder [*numerator*, *denominator*, *variable*]

Returns the continued fraction expansion of an impedance or admittance. The numerator, denominator, and complex frequency variable must be given.

ChebyP[n,w]

Returns the Chebyshev polynomial of order n and variable w. The results are identical to **ChebyshevT** in *Mathematica*. **ChebyP** illustrates a recursive function definition.

Circuit[..]

Is a wrapper which holds **Conductance** circuit components separated by commas.

Components[*filter*]

Returns the normalized component values (g_i) for the given *filter*. Termination values are included. *filter* may be **Butterworth**[n = **order**] or **Chebyshev**[n = **order**, e = **ripple**].

Conductance[{*node1*,*node2*},*value*]

Defines a circuit admittance component for **YParams**.

Cor[..]

A function that indicates correlation between two variables. The only action **Cor** takes is to simplify the sum of several signals into the sum of correlations. **Cor** is used for algebraic simplification in the symbolic analysis of noise.

Correlate[*x1*,*x2*:**optional**]

Returns the correlation of *x1* with *x2*. If only *x1* is given then the autocorrelation of *x1* is performed.

cost[*Inductance*,*Length*,*Diameter*]

Returns the relative cost in cents for a solenoid of given inductance, length, and diameter. Requires *Nodal*.

CrossPowerSpectrum[*x1*,*x2*]

Returns the cross-power spectral density of *x1* and *x2*.

DrawCascade[**Cascade**[..]]

Draws the block diagram for the system cascade. **Cascade** is a wrapper that holds one or more **Amplifier** components.

error[*C1***,***L2***,***frs***,***mags***]**

Returns the squared error for the high-pass filter optimizer in Chapter 8. The filter elements are given by *C1* and *L2*. The test frequencies are given by *frs* megahertz. The goal magnitudes at each frequency are given by *mags*. Requires *Nodal*.

FilterForm[..]

Is the wrapper for filter component data.

GainCircle[*Gp***,{{***s11***,***s12***},{***s21***,***s22***}}]**

Returns the gain circle center and radius for a power gain of G_p and the given device S-parameters.

h2B[*n***,***s***]**

Returns the squared magnitude of the Butterworth filter transfer function:

$$\texttt{1/(1 + (-}s\texttt{\textasciicircum2)\textasciicircum}n\texttt{)}$$

h2C[*n***,***s***,***e***]**

Returns the squared magnitude of the Chebyshev filter transfer function:

$$\texttt{1/(1 + }e\texttt{\textasciicircum2 ChebyP[}n\texttt{,}s\texttt{/I]\textasciicircum2)}$$

HammingWindow[*i***,***points***]**

Returns the Hamming window coefficient for position *i* in an array of length *points*.

HtoZin[*S21Squared***, ***variable***]**

Returns the input impedance of the network with squared magnitude transfer function *S21Squared*. The complex frequency variable must be given as *variable*. A doubly terminated network is assumed.

ListSum[*list***]**

Returns the sum of the list.

MAG[*s21m***,***s12m***,***K***]**

Returns the maximum available gain for a device with the given S_{21} magnitude, S_{12} magnitude, and *K* factor.

MSG[*s21m, s12m***]**

Returns the maximum stable gain for a device with the given S_{21} magnitude and S_{12} magnitude.

MUG[*s21m, s11m, s22m***]**

Returns the maximum unilateral gain for a device with the given S_{21} magnitude, S_{11} magnitude, and S_{22} magnitude.

PlotCascade[Cascade[..], *options***]**

Plots the system analysis of the cascade — the results of **Tabulate**. **Cascade** is a wrapper that holds one or more **Amplifier** components. The *options* are *Mathematica* **Graphics** options.

PolePlot[*function, variable***]**

Plots the pole locations of *function* for the given *variable*.

PowerSpectrum[*data***]**

Returns the power spectrum of the list *data*.

rho[*z, zo***:50]**

Returns the reflection coefficient for the impedance *z* in system impedance *zo*. The default *zo* is 50.

SD[..]

SD is short for spectral density. **SD** works with the **Cor** function to help in symbolic analysis of noise signals. **SD** sets correlations of different signals to zero and correlations of the same signal to a spectral density of that signal.

Sinc[*x***]**

Defines a **Sin[***x***]**/*x* function with a value of 1 for **Abs[***x***]** < **10^-9**.

SmithCharter[*rhos, options***]**

Shows a Smith chart and plots the given points on the chart. *rhos* can be zero or more lists of complex reflection coefficients. The *options* should be *Mathematica* **Graphics** options.

smithTraceY[*zAdded***,** *zOld***,** *points* **: 5]**

Returns {list of reflection coefficients, final Z}. The reflection coefficients come from points between *zOld* and the parallel combination of *zOld* and *zAdded*. The default number of points is 5. The final impedance is *zOld* in parallel with *zAdded*.

smithTraceZ[*zAdded***,** *zOld***,** *points* **: 5]**

Returns {list of reflection coefficients, final Z}. The reflection coefficients come from points between *zOld* and *zOld* **+** *zAdded*. The default number of points is 5. The final impedance is *zOld+zAdded*.

solenoidFacts[*length***,** *numTurns***,** *diameter***]**

Prints the solenoid diameter in microns, total wire length in meters, and resistance in ohms.

SolveCircuit[YData[..], *currentsList***]**

Returns the node voltages when the currents in *currentsList* are applied to the matrix in **YData**. **YData** is generated by **YParams**.

sPlot[σ, ω**]**

Plots the Complex plane, real axis and imaginary axis projection of a signal generated from the given pole, $\sigma + I \, \omega$.

StabilityCenter[{{*s11***,** *s12***},{***s21***,** *s22***}}]**

Returns the center of the stability circle for the given S-parameters.

StabilityCircles[{{*s11***,** *s12***},{***s21***,** *s22***}}]**

Returns rules for the center and radius of the input and output stability circles for the given S-Parameters.

StabilityK[{{*s11***,** *s12***},{***s21***,** *s22***}}]**

Returns the stability factor, *K*, for the given S-parameters.

StabilityRadius[{{*s11***,** *s12***},{***s21***,** *s22***}}]**

Returns the radius of the stability circle for the given S-parameters.

szMap[*s***]**

Returns a {magnitude,angle} pair equal to **Exp[***s***]**. Requires *Nodal*.

Tabulate[Cascade[..]]

Returns a table of system performance (gain, NF, IP3) as each component of the system is added. The **Cascade** wrapper holds one or more system components (**Amplifier**) separated by commas. The IP3 is referred to the system input.

Transform[*comps***,***Zo***,***startComp***,***wo***,***type***]**

Returns a table of filter element values as a **FilterForm** data type. Parameters include a list of component values which begin and end with source and load resistances, the system impedance, *Zo*, the starting component for the low pass equivalent (**L** or **C**), the radian frequency for scaling, and the transformation type (**LP**, **HP**, or **BP[Qo]**). The band pass type should include a number for **Qo**.

yLoad[SMatrix[..], *fr***]**

Returns a **YMatrix** data type with the **Frequency** option set to *fr*. This is used to load S-parameter data into the *Nodal* **SubNetwork** component.

YData[..]

Is a wrapper for Y matrix data. **YData** works just like **YMatrix** in *Nodal* but a different name was used to avoid conflicts.

YParams[ABCDData[..] or Circuit[..]]

Returns the *Y*-parameter matrix (as **YData**[..]) for the given matrix data, such as **ABCDData**, or for the given **Circuit** data. **YParams** works like **YParameters** in *Nodal* but a different name was used to avoid conflicts.

zFilterPZ[*data***,{***p***,***z***}]**

Returns *data* filtered by the *z*-domain pole, *p*, and zero, *z*.

zFromRho[*rho***,** ***zo***:50]**

Returns the impedance that creates reflection *rho* in system impedance *zo*. The default *zo* is 50.

A.7 Complex Algebra

Manipulating and simplifying algebraic expressions is one of the most useful features of *Mathematica*. However, at times the result is returned in a strange form or takes an unexpectedly long time to produce. The **Re**, **Im**, and **Conjugate** functions in the **ReIm.m** package work exactly as they should, but in some cases take extremely long amounts of time to do seemingly simple manipulations. In Section 4.6 of this book you analyzed power transfer between complex source and load impedances. Your problem was to find the real power in the load.

```
In:
  Zs = Rs + I Xs;
  ZL = RL + I XL;
  Current = Vs/(ZL+Zs);
  VL = Current ZL;


  Needs["Algebra`ReIm`"];
```

According to the **ReIm** package information you should set the imaginary part of the variables to zero when the variables are real:

```
In:
  Rs /: Im[Rs] = 0;
  Xs /: Im[Xs] = 0;
  RL /: Im[RL] = 0;
  XL /: Im[XL] = 0;
  Vs /: Im[Vs] = 0;
```

The solution is repeated below with the timing information given for a Macintosh IIcx computer:

```
In:
  (tmp = Re[VL Conjugate[Current]];) //Timing
Out:
  {5291.65 Second, Null}
```

The solution takes a long time because of the way **ReIm** works. By removing the variables and resetting the equations you can see how **ReIm** works. **ReIm** is still loaded. The equations below solve for the real part of **Current**. Since the tests setting the imaginary part of each variable to zero was removed we get a large expression with **Re** or **Im** wrapped around each variable. **ReIm** works by breaking down the input expression until it finds a single variable to test. This results in "expression swell" and consumes a large amount of time.

```
In:
  Remove[Rs,Xs,RL,XL,Vs];
```

```
Zs = Rs + I Xs;
ZL = RL + I XL;
Current = Vs/(ZL+Zs);
VL = Current ZL;
```

Current

Out:

```
            Vs
--------------------
RL + Rs + I XL + I Xs
```

In:

tmp = Re[Current]

Out:

```
((-Im[XL] - Im[Xs] + Re[RL] + Re[Rs]) Re[Vs]) /
                                                   2
    ((-Im[XL] - Im[Xs] + Re[RL] + Re[Rs])  +
                                                   2
     (Im[RL] + Im[Rs] + Re[XL] + Re[Xs]) ) +

   (Im[Vs] (Im[RL] + Im[Rs] + Re[XL] + Re[Xs])) /
                                                   2
    ((-Im[XL] - Im[Xs] + Re[RL] + Re[Rs])  +
                                                   2
     (Im[RL] + Im[Rs] + Re[XL] + Re[Xs]) )
```

For typical electrical engineering problems there is a better way to manipulate complex expressions. You basically have two kinds of complex expressions: those with complex values in the denominator and those with only real values in the denominator. Any expression with complex values only in the numerator is reduced to its real part by substituting **Complex[u_,v_] :> u** throughout the expression. The following function, **getReal**, does this under the condition that the denominator is free of complex expressions. You must use **Complex**, rather than **I**, to match complex expressions. All of the following functions assume any symbol other than **I** is real. You could also use **ComplexExpand** to simplify complex expressions. However, using **ComplexExpand** as a foundation for **getReal** and **getImag** produced longer analysis times than the following functions, and did not always remove the **I** terms.

In:
```
getReal[x_] :=
    Module[{u,v},
        x /. Complex[u_,v_] :> u
    ] /; FreeQ[Denominator[Together[x]],Complex]
```

Next you need to convert any complex expression into an expression with a real denominator. The following function, **getNumDen**, returns the numerator and denominator of any expression.

In:
```
getNumDen[x_] :=
    Module[ {tmp},
        tmp = Together[ExpandAll[x]];
        Expand[{Numerator[tmp],Denominator[tmp]}]
    ]
```

Now all you have to do is realize that:

$$\frac{x + Iy}{u + Iv} = \frac{x + Iy}{u + Iv} \frac{u - Iv}{u - Iv}$$

$$= \frac{(x + Iy)\ (u - Iv)}{u^2 + v^2}$$

Once you find the real part, the imaginary part, and the complex conjugate of a complex denominator you can easily construct an expression with only a complex numerator. The latter expression is solved by our previous form of **getReal**. The following function defines **getReal** for those expressions with complex values in the denominator:

In:
```
getReal[x_] :=
    Module[{n,d},
        {n,d} = getNumDen[x];
        getReal[Expand[n getConj[d]]]/
            (getReal[d]^2 + getImag[d]^2)
    ] /; !FreeQ[Denominator[Together[x]],Complex]
```

The next set of functions define **getImag** and **getConj** similar to how you defined **getReal**:

```
In:
  getImag[x_] :=
  (
      getReal[ExpandAll[-I x]]
  ) /; FreeQ[Denominator[Together[x]],Complex]


  getImag[x_] :=
      Module[{n,d,u,v},
          {n,d} = getNumDen[x];
          getImag[Expand[n getConj[d]]]/
              (getReal[d]^2 + getImag[d]^2)
      ] /; !FreeQ[Denominator[Together[x]],Complex]


  getConj[x_] :=
      Module[{u,v},
          x /. Complex[u_,v_] :> Complex[u,-v]
      ] /; FreeQ[Denominator[Together[x]],Complex]


  getConj[x_] :=
      Module[{n,d,u,v},
          {n,d} = getNumDen[x];
          getConj[Expand[n getConj[d]]]/
              (getReal[d]^2 + getImag[d]^2)
      ] /; !FreeQ[Denominator[Together[x]],Complex]
```

Once **getReal**, **getImag**, **getConj**, and **getNumDen** are defined you
should revisit the problem from Section 4.6. The solution is given below as timed
on a Macintosh IIcx. We have improved the function execution speed by over 3
orders of magnitude. The expression in **realPower** can be reduced further by
Simplify. If you use the **ReIm** package you will initially get a much more complicated expression for **realPower** (because of the way the package works);
however, **Simplify** will reduce each result to the same expression.

```
In:
  realPower=getReal[VL getConj[Current]]//Timing
```

```
Out:
                              RL Vs (RL Vs + Rs Vs)
  {3.21667 Second, ------------------------------------}
                                        2          2
                   (RL + Rs) ((RL + Rs)  + (XL + Xs) )
```

Index

(*
  for comments 17
;
  separating commands 17
  to suppress output 17
->
  toolbox 16, 321
?
  obtaining help 14
  to determine assignments 15
_ 252
  in names 14
π-pad
  design 41

1 292
1/f noise 292

ABCD-Parameters
  definition 102
Absolute value
  Abs function 62
Admittance 324
Aliasing
  frequency of sampling 232
  sampling resulting in 231
Amplifier 276
  charge-sensitive 89
Amplifiers
  operational 83
Analog filters
  band-pass 206
  component values 196
  doubly terminated synthesis 216
  frequency scaling 201
  high-pass 204
  impedance scaling 199
  pole-zero locations 193
  singly terminated synthesis 211
  transfer functions 189
Analysis
  stability 254
Anonymous functions 323
  toolbox for 108
ArcTan
  toolbox 247
Arg 62

Argument
  Arg function 62
arma 237
Assigning
  names 14
Assignment 10, 320
Assignments
  inquiring about with ? 15
  with rules 16, 321
Attenuator
  L design 42
Attenuators
  design of 41
Autocorrelation 295
AWG
  designing with 143

Bar charts 155
Bias point
  FET 54
  for diode 52
Bilinear transform 234
Bipolar transistor 325
  model 91
BJT 325
Bode plot 80
  oscillator 96
  RLC filter example 136
Bridged-T
  transistor model 91
Brown noise 290
  generating 290
Butterworth
  design 225
  filter response 191
  transfer function 189

Capacitance
  FET model 91
  stray 65
Capacitor
  calculating value of 133
Capacitors 325
cascade analysis 274
CCVS 89
Chain parameters
  definition 102

Chart
  bar 155
  error bars 155
Charts
  pie 155
ChebyP 192
Chebyshev
  filter response 192
  ripple parameter 192
  transfer function 189
Chebyshev polynomials 170
Circle
  toolbox 246
Circuit analysis
  current through node 32
  frequency specification in 88
  h-parameters in 75
  linear and nonlinear 74
  multistate DC circuits 33
  nonlinear circuits 43
  power supplies in 75
  small signal 74
  transfer function 173
  using matrices 30
  voltage at node 32
Circulator 325
Clearing variables 10, 320
Comments
  inserting 17
  syntax for 319
Comparing graphs
  using Show 140
Complex numbers 321
  toolbox 63
Components 197
  specifying to *Nodal* 9
Conversion factors
  in *Nodal* 13
Correlation (noise) matrices 300
  combining 307
  generating 300
Cost minimizing 148
CoupledInductors 326
Current-controlled sources 325
Current-controlled voltage source 89
Current sources
  impedance of 75

voltage-controlled 85
CurrentNoise 326
CurrentSource 326
CVal 227

Data
    fitting functions 47
    manipulating read-in 156
    reading from files 44
    type specification 45
dB
    expressing ratios in 61
Deassignment 10, 320
Decibels
    expressing ratios in 61
Denominator
    toolbox 94
DensityPlot 204
Derivatives
    sensitivity analysis using 145
Differential sensitivity analysis 145
Differentiation
    toolbox 78
Digital filters 230
    bilinear transform 234
    design by pole-zero placement 235
    finite impulse response 238
    infinite impulse response (IIR) 233
    realizing 236
    transfer function 235
    unit circle 233
Diode 326
    calculating bias point 52
    characteristics 51
Distortion 275
Dump
    to save files 6

Enumerating (N) 29
Equations
    solving sets of 22
Error 228
Error function
    in filter design 228
Error messages
    controlling 18
    generally 18
    nodeOrder 10
Error bars
    plotting 155
Errors
    reporting with Message 262
Exponentiation 12

Faster start up
    using Dump 6
FET
    calculating bias point 54
    capacitances in model 91
    characteristics 54
    transconductance 91
FET model 90
FETs 326
    generally 85
Field Effect Transistors
    see FETs 85, 326
Filter
    Sallen-Key 173
Filters
    designing 224
    digital 230
    frequency-domain 167
    specification 225
FindMinimum 229
FindRoot 278
Finite impulse response
    digital filter type 238
Fitting functions 47
FoldList 197, 290
Format 276
    number of digits in N 29
Formats
    numbers 320
Formatting
    with N 154
Fourier 239, 295
Fourier analysis 165
Fourier transform
    noise 288
FourierEE 239
Frequency
    oscillator 94
    sampling 232
Frequency specification
    in circuit analysis 88
Friis 274
Functions
    anonymous 108, 323
    fitting to read-in data 47
    introducing user-defined 37
    synthesizing 163
    user-defined 322

Gain
    and transconductance 85
    circles 260
    falloff with frequency 79

maximum available 260
maximum stable 260
maximum unilateral 260
Gaussian 301
Get
    to load Nodal 5
Graph
    pie 155
    polar 155
    scatter with ListPlot 154
Graphics
    simultaneous display using Show 140
Graphics primitives 246
    toolbox 246
GraphicsArray 249
Gyrator 326

Hamming 239
Help
    argument template for functions 15
    finding full names 15
    obtaining on-line 14
Histogram
    of signal 287
h-parameters
    bipolar transistor model 92
    transistor circuits 75
Hysteresis 33

Identifying
    nodes 9
Impedance
    current sources 75
    transforming parameters 128
    voltage sources 75
Impulse response 160
    calculating 176
Inductor 327
    design example 136
Infinite impulse response
    digital filter type 233
Installing
    Mathematica 5
    Nodal 5
Intercept point 273
    IP3 275
Inverting matrices 31
Isolator 327

JFET 85

K
    Rollett stability factor 254

Kirchoff's laws 6, 26

L attenuator
  design 42
Labelling points 155
Laplace transforms
  inverse finding 179
Limit 82
Line
  toolbox 246
Linear circuits
  caveats 7
Linear series
  generating with Range 88
List
  generation using Table 153
ListPlay 237, 289
ListPlot
  scattergraph creation 154
Lists
  generally 320
  generating with Range 88
  introducing 8
  making functions process 261
  taking parts of 47
ListSum 277
Loading
  Nodal 5
  using Needs 5
Loading packages 155
Logarithmic series
  generating with Range 88
Logarithms
  toolbox 47

MAG 260
Map 237, 248, 293
Mapping
  impedance to reflection 246
Matching networks
  design of 265
Mathematica
  installing 5
Matrices
  correlation 300
  solving DC networks 30
Matrix
  definite and indefinite admittance 106
Matrix inversion 31
Matrix methods 30
Matrix transposition 31
Mean
  of noise 287

MESFET 85
Message function 262
Miller effect 82
Minima
  finding 53
Missing nodes 10
Monte Carlo
  inverse functions for 160
MOSFET 85
MSG 260
MUG 260
Multiplication
  of matrices 30, 322

N
  formatting 154
  precision specifying 29
Names
  underscore in 14
Naming variables 10, 320
Needs
  to load *Nodal* 5
Negative resistance 77
Netlist
  conversion for matrix calculation 105
  introducing 7
Networks
  shunt 267
Nodal 271
  installing 5
  loading 5
Nodal Admittance Matrix
  s-domain version 173
NodalAnalyze 227, 332
  introduction 24
NodalNetwork
  introduction 24
NodalPlot 229
Nodes
  identifying numbers 9
  missing 10
Noise
  1/f 292
  analysis with *Nodal* 315
  brown 290
  creating white 286
  description of 285
  generating time-series 157
  histogram comparison 295
  matrix analysis 304
  mean 287
  multiple sources 301
  pink 292

resistor 300
  variance 287
  white 286, 287
Noise figure 273, 310
Nonlinear circuit analysis 74
Nonlinear circuits
  caveats 7
Nonlinearity 273
Normal 202
Norton equivalence
  for current sources 106
Not 224
Numbers
  complex 321
  formats 320
  random 323
  syntax 320
  toolbox for complex functions 63
Numerator
  toolbox 94

Ohm's Law
  solving 21
Operational amplifier
  model for 83
  transfer function specification 83
Operational amplifiers 83, 328
Operator syntax 322
Optimization 224
Options
  introducing 16
  passing to other functions 253
  using 16
Oscillator
  Bode plot 96
  relaxation 177
Oscillators
  criterion for oscillation 93
  frequency of oscillation 94
  generally 93
Outer 300
Output
  suppressing 17
Overloading
  in functions 190

Packages
  loading 155
Pad
  L design 42
Pads
  design of 41
Parameter calculation 331

Parameters
  introducing 100
Partial derivatives
  sensitivity analysis using 145
Partial fractions
  in *s*-domain problems 179
Phase 61
  Bode plot of 136
Phase-shift
  Bode plot 96
  oscillators 93
Pie charts 155
Pink noise 292
  generating 292
PlotCascade 279
Plots
  avoiding singularities in 190
Plotting
  polar graphs 155
  range restricting 136
Plotting function
  Bode plot 80
Point
  toolbox 246
Point spread function 160
Points
  labelling 155
PointSize
  toolbox 246
Poisson inter-event times 170
Polar coordinates
  for complex numbers 63
  s- to *z*-plane mapping with 233
Polar plots 155
PolePlot 195
Poles
  locating 174
Pole zero
  locations of 174
Pole-zero extraction 220
Pole-zero visualization 174
Polynomial
  manipulation 324
Power 295
  measurement in correlation 298
Power spectral density 296
Power supplies
  in linear circuit analysis 75
Precision
  specifying with N 154

Primitives
  graphics 246
Probability density function
  of noise 285

Q
  as circuit parameter 70
  in analog filters 207
  in resonant circuits 183
  resistance effects 144
Qualifiers
  list of in *Nodal* 12

Random
  toolbox 157
Random numbers 323
  prescribed distributions 157
Range
  generating log series with 88
  toolbox 88
Ratio
  expressing in dB 61
Reflection coefficients 258
  calculation of 245
Reserved variables 10
Resistance
  negative 258
  negative dynamic 77
  skin effect at RF 142
  specifying resistivity 142
  static and dynamic 76
Resistivity
  specifying 142
Resistor 328
  noise in 300
Resonance
  in high Q circuit 71
Rollett stability factor 254
  with S-parameters 255
Rolloff
  in filters 191
Root finding
  inductor design example 137
Roots
  finding 53
Rules
  toolbox 16, 321

Sampling
  aliasing caused by 231

frequency of 232
Saving files
  using Dump 6
Scaling values
  using qualifiers 11
Scattergraph
  with ListPlot 154
Schmitt trigger 33
s-domain
  nodal admittance matrix 173
  pole-zero visualization 174
Selectivity
  resistance effects 144
Sensitivity analysis 138
  differential 145
Sequences
  prescribing with underscores 252
Series
  logarithmic 88
  using Range to generate 88
Signals
  description in *s*-plane 171
Sinc 190
Singularities
  dealing with in plots 190
Skin effect 142
Smith chart
  admittance version 267
  function 250
  generating 248
  introduction 245
  stable regions 254
SmithCharter 252
smithG 252
Solve command 22
S-parameters
  stability checking with 258
s-plane
  signal description 171
spreadsheet 273
Stability
  checking with S-parameters 258
  circle calculation 258
  conditional 256
  Rollett factor 254
Stability analysis 254
Step response
  calculating 176
Stray capacitance 65
SubNetwork 272

Syntax
    comments 319
    generally 319
    numbers 320
    operators 322
szMap 233

Table
    list generation using 153
    time-series generation using 153
Tabulate 276
Templates 15
Text
    toolbox 246
Time series
    generally 153
    generating 153
    generating with Table 153
Time-domain behavior
    from s-domain 176, 181
Toolbox
    % 25
    anonymous functions 108
    ArcTan 247
    arithmetic operators 11, 321
    assigning and deassigning 10, 320
    Circle 246
    complex number operations 63
    Denominator 94
    differentiation 78
    exponentiation 12
    finding minima 53
    finding roots 53
    fitting functions to data 47
    general command syntax 8
    graphics primitives 246
    infix 28
    introducing 7
    Line 246
    lists 8
    logarithms 47
    matrix methods 30

N 29
Numerator 94
Point 246
PointSize 246
postfix 28
prefix 28
Random 157
Range 88
reading in data 45
recalling last answer 25
replacement 28
rule(->) symbol 16, 321
Simplify 25
Solve 22
taking parts of lists 47
Text 246
user-defined functions 37
T-pad design 41
Transconductance
    calculating 85
    FET circuit 91
Transfer function
    in s-plane 173
    operational amplifier 83
    steady-state 182
TransferDelay 329
TransferFunction 329
TransferPhase 329
Transform 206
Transformer 329
Transistor
    bipolar 325
TransmissionLine types 330
Transposing matrices 31
Trigger
    Schmitt 33
Tunnel diode
    dynamic and static resistance of 77

Underscore
    in names 14
    usage 252

Unit circle
    digital filters and 233
    z-plane and 233
Units
    use of SI/mks 11
Unset
    introducing 10, 320
User-defined functions 322
    introducing 37

Variables
    clearing values assigned 10, 320
    eliminating in Solve 24
    naming 10, 320
    reserved 10
Variance
    of noise 287
VCCS
    generally 85
VCVS 79
Voltage 75
Voltage-controlled current sources 85
Voltage-controlled sources 330
Voltage-controlled voltage sources 79
Voltage source
    current-controlled 89
    impedance of 75
    voltage-controlled 79
VoltageSource 331

Wrapper 276

Y-parameters
    definition 100

Zeros
    locating 174
Z-parameters
    conversion to S 127
    definition 101